CITY SLUMS

INEQUALITY IN THE CITY

CITY SLUMS

A Political Thesis

J.A. INGHAM

Routledge
Taylor & Francis Group

LONDON AND NEW YORK

First published in 1889

This edition published in 2007
Routledge
2 Park Square, Milton Park, Abingdon, Oxon, OX14 4RN

Simultaneously published in the USA and Canada by Routledge
711 Third Avenue, New York, NY 10017
Routledge is an imprint of Taylor & Francis Group, an informa business

Transferred to Digital Printing 2007

First issued in paperback 2013

© J.A. Ingham, Jun

The publishers have made every effort to contact authors and copyright
holders of the works reprinted in the *The City* series. This has not been
possible in every case, however, and we would welcome correspondence
from those individuals or organisations we have been unable to trace.

These reprints are taken from original copies of each book. In many cases
the condition of these originals is not perfect. The publisher has gone to
great lengths to ensure the quality of these reprints, but wishes to point out
that certain characteristics of the original copies will, of necessity, be
apparent in reprints thereof.

British Library Cataloguing in Publication Data
A CIP catalogue record for this book
is available from the British Library

City Slums

ISBN13: 978-0-415-41827-0 (volume hbk)
ISBN13: 978-0-415-41926-0 (subset)
ISBN13: 978-0-415-41318-3 (set)
ISBN13: 978-0-415-86048-2 (volume pbk)

Routledge Library Editions: The City

CITY SLUMS

LAND DEPRESSION AND COMMON LABOUR
" THE ANOMALY OF LEGISLATION "

A POLITICAL THESIS

BY

J. A. INGHAM, Jun.

WITH

A REVIEW OF VISCOUNT CROSS AND TORRENS ACTS

BY

J. GORDON M^cCULLAGH, ESQ.

BARRISTER-AT-LAW, TEMPLE

LONDON
SWAN SONNENSCHEIN & CO
PATERNOSTER SQUARE
1889

CONTENTS.

Preface.

SINCE MR. GODWIN, in 1854, wrote "London Shadows," the questions of adequately providing fit dwellings for the poor of that vast Metropolis, and our utilization of sewage for agricultural purposes, which otherwise becomes not only a waste but an incubus, have each passing year occupied a more important place in the field of domestic Legislative Science. The former in the judgment of the analytical jurist, who is also a practical politician, merits a foremost place in English thought, so important indeed, that it cannot well be overrated.

Eminent lawyers and great authorities have declared that English jurisprudence is based on the principles of Christian morality; as otherwise our legislators could never have succeeded in assisting it to give to her people the softening and elevating influences of a true religion, education, and refinement; and are not these characteristics also a marked feature in her literature?

In submitting this work to the public it must
be distinctly understood that it is written im-
partially and without party bias. There are
those members of the English bar, who, reared
at the feet of the great Gamaliels of the past,
have made our legislative science, dry subject of
research as it is, almost a life-long study ; and if
as they read, some should in good earnest take
up the question, with more thoroughness than
this pen is able to accomplish, and some amount
of good result ensue, the object in view will
have been attained.

The endeavour has simply been to deal with
the broad question, and it is feared, that, only in
a fragmentary way.

The fact being well ascertained that the
dwellings of the London poor are in the highest
degree unhealthy and greatly overcrowded ;
let us strive to prove, leaving provincial corporate cities and incorporated towns to deal with
their own evils, that State interference as regards
the former, and subsequent surveillance as to
overcrowding, can alone grapple with this
immense evil in the Metropolis.

PART I.

CHAPTER I.

The Anomaly of Legislation.

NOW that State directed colonization and English Land Nationalization are occupying the minds of thinking men, among whom are to be found some of the most eminent politicians of the day; should some such system as the former be adopted, the large masses of our population, in seeking the virgin fields of the far Western States of America, would but be following the tide of civilization, as seen in the course which providence ordained the sun to take, rising in the East and setting in the West.

But is such a scheme politic ? Have we not all a right to remain in the land of our birth ?

Amongst other things, want of employment in rural districts, during comparatively recent years, has greatly tended to increase the congestion of the country population in our large towns, especially the Metropolis ; for the country workman has ever the idea that the great centres of labour always are the best field for work, whereas, the opposite is almost invariably the case.

Without, however, dealing with London exclusively at present, let us turn to a few statistics in order to appreciate the importance of this question of overcrowding, which is now, alas, a national one, with the object of seeing whether State directed colonization and English Land Nationalization are the only remedies.

CHAPTER II.

Statistics.

THE Local Government Chronicle of the 7th August, 1875, tells us that in a certain area bordering upon Gray's Inn Road there were 239 houses containing 1,019 families, making an average of four families to a house. Some of the alleys in which this multitudinous population lives are but four feet wide, the district being one of the most densely crowded quarters of London. This periodical, of the same month and year, also tells us that the local Medical Officers of Health of the Whitechapel and Holborn districts, reported a neighbourhood having a total area of only eleven acres, in which there was one block of seven houses containing forty-two families, thus averaging six families to a house, and containing a total number of 168 people, of whom (assuming a fairly equal distribution) there were twenty-four people to each house. On this same area of eleven acres was another block of fifteen houses, in which there were housed ninety families composed of 360 persons, which would also allow about twenty-four people to each house. In a third block, on a similar area of eleven acres, there were nineteen houses occupied by 101 families (five families to a house) composed of 608 souls, averaging thirty people to each house. The parts of London, including the city proper, and the vast range of houses and streets within the county of Middlesex, where this great evil especially exists, are the Euston Road, the Old Ford, St. George's in the East, the Ingestre Buildings, the area surrounding the slaughter-houses at Whitechapel, the narrow courts and tortuous lanes of Smithfield, and Anne Street and Pye Street, Westminster, the Brill at Somer's Town, certain parts of Lisson Grove, and the worst parts of the parishes of Lambeth and St. George-the-Martyr. The districts of Clerkenwell, St. Luke's, Holborn, Bermondsey, St. Mary's, Newington, and Whitechapel are perhaps those which ought to be first attended to. The Local Government Board had communicated to the St. George's (Southwark) Vestry a report from Dr. Ballard, one of their Medical Inspectors, calling attention to some of the miserable and unhealthy dwellings in that

district, the sanitary officers of which, a short
time before, appear to have denied that there
were any such objectionable cases within their
knowledge. Dr. Ballard describes these same
dwellings which are called " Fisher's Buildings, "
" as being a row of seven small houses in a court
" entered from the north side of George Street,
" Tabard Street. From a tablet upon one of
" them, it appeared that they were erected in
" 1821, at a time when many similar small houses
" built with inferior materials were run up in the
" outskirts of the City. The houses face north-
" wards, and have small courtyards in front of
" them, and these, with the footway, are badly
" paved with bricks. Each house consists of a
" sitting room below, and a bedroom above.
" They have no back openings. They are all
" very dirty and much dilapidated, and the walls
" are damp from the foundations, due, probably,
" in great part, to the inferior and absorbent
" quality of the bricks used in their construction.
" No amount of repairs would render these
" houses wholesome habitations, and in my
" opinion they should have been dealt with under
" the Artisans' and Labourers' Dwellings Acts.
" The article goes on to state that the Medical
" Officer of Health of the district had also

" reported against many similar dwellings, but it
" may be added only at the eleventh hour.
" These houses were stated by a member of the
" vestry to be a standing danger to the health of
" the parish, and, beyond all doubt, a disgrace to
" a civilized community."

The infant mortality is appalling !—It would
appear from a comparison, also given in the
Local Government Chronicle of the 27th of
February, 1875, of the death rate amongst
infants of the Hyde Park districts, with that of
those living in an equal area of the East End,
that, where twelve children grew up to maturity
in the former only one became adult in the
latter. It must, however, be admitted that the
death rate generally has improved during the
last twenty years ; still it is very far from what
it ought to be. And from the same periodical,
of the 26th of June, 1875, from the report of
the proceedings in the House of Lords, on the
22nd of that month, the late Lord Henniker
ascertained the proportion of rural population
which had migrated to towns during a period of
ten years with the following results which are
most important. The population in Lincoln-
shire, in 1861, was 404,138 ; in 1871 it was
428,075 ; but under ordinary circumstances the

natural increase would have been 56,060 ; instead of which it was only 23,937, leaving a deficiency of 32,123 in the ten years' time. The population in Suffolk in 1861 was 335,409 ; in 1871 it was 347,210, its increase was only 11,801, but under ordinary circumstances it ought to have been 41,250, thus leaving a deficiency in the rural population of that county of 29,449, in the decennial period. Now we will contrast the above particulars with the state of things in London. The population in 1861 was 2,803,989 ; in 1871 it was 3,254,260 ; the natural increase should have been 321,870, instead of which it was 450,271, implying a surplus of 128,401 within the decade, and in 1881 its population was 3,813,000, equal to that of the whole of Scotland. We will now take the great central North country town of Leeds. Its population in 1861 was 134,006 ; in 1871 it was 162,421, the ordinary and natural increase being 18,580, instead of which it was 28,415, *i.e.*, 9,835 more than it ought to have been. It is computed that about 2 millions leave the rural districts for the towns in but ten years time in England alone. The evil is National ! How can it be arrested ? It must be dealt with by the Patriotic and zealous workers ! by

those who have a universal love for Humanity!
Not by the pleasure seeking indifferent shirkers.

The Poor Law Amendment Act, 1825, enables the guardians to purchase 50 acres and let to the poor cottagers for cultivation, evidently to prevent vagrantcy and outdoor relief?

This Act is now being utilised extensively in many rural districts, the cultivators generally living in the workhouse, and the produce is sold by the guardians. The object manifestly is that each inmate should by his or her labour make a return towards a reduction of the rates! This is a most hopeful sign of the times!

If this principle is applicable in a local and provincial point of view by Poor Law Authorities throughout the United Kingdom, why should not the more universal arable cultivation of the soil be the most available outlet for healthy honest labour preventing poverty instead of attempting to cure it when it has once grasped its victims?

Surely this is a problem which we cannot too speedily solve! From the above statistics it must be quite clear, that even, with a merely normal increase of population, London would be very much overcrowded, quite apart from the fact that our rural population has for some years past been flocking into it. Now this overcrowding is not of recent origin in London, for the state of things there was bad enough three hundred years ago. Sir Lyon Playfair once traced it from the days of Queen Elizabeth, and showed its periodical growth up to the present time, and the evil would appear to have more causes than one.[1]

[1] From the columns of the *Daily News*, which has now for some years given this subject a prominence, it appears that Liverpool in parts is more thickly populated than London; but its sanitation is much better. Lord Basing's address on the 20th September, 1887, to the Sanitary Congress is a review of progress and it alleges that it is possible to reduce the Metropolitan death rate from 19·3 in the 1000 which it was at in 1885 to 5 in the 1000.

CHAPTER III.

𝔏𝔬𝔫𝔡𝔬𝔫 𝔖𝔩𝔲𝔪𝔰 𝔞𝔫𝔡 𝔱𝔥𝔢𝔦𝔯 𝔒𝔠𝔠𝔲𝔭𝔞𝔫𝔱𝔰.

IN order to get into the dwellings of these unfortunates, *Les Misérables* allowed by our modern civilization, we have to penetrate into courts and alleys reeking with poisonous gases, arising from bad drainage and accumulation of refuse scattered in all directions—courts, many of them, into which the sun never penetrates, which are never visited by a breath of fresh air, and which rarely know the healing virtues of cleansing water. We have to ascend decayed staircases which threaten to give way beneath every step, and which in some places have already broken down, leaving gaps which imperil the lives and limbs of the unwary. We have to continue groping our way along dark and filthy passages swarming with vermin, before we gain admittance into the dens in which thousands of human beings herd together, who belong as much as ourselves to the race for whom legislation and Christianity have claimed to labour, and fight, and upon whom to confer an even handed justice.

The walls and ceilings are black with the accumulation of filth, which has gathered upon them through long years of neglect. What goes by the name of a window is generally half of it stuffed with rags or covered by boards to keep out the wind and rain, and what little glass there is, is so begrimed and obscured, that light can scarcely enter, nor from within can anything of the outside be seen. Should we ascend to the attic, where at least some approach to fresh air might be expected to enter from a broken or an open window, we look out upon the roofs and ledges of lower tenements, and discover that the sickly air which finds its way into the room, has to pass over the putrefying remnants of dead birds, or cats, or viler abominations still. The buildings are in so ruinous a state, as to suggest the suspicion, that if the wind were not intercepted by other outside buildings, the roofs would fall in upon the heads of the occupants.

As to furniture, we may perchance discover a broken chair, and the mere fragment of a table, but more commonly we shall find rude substitutes for these in the shape of rough boards resting upon bricks, or upon an old hamper, or a box turned upside down, or more frequently still, nothing but rubbish and rags. Every room in these decayed buildings houses a family, often two. In one cellar, a sanitary inspector reported finding a father, mother, three children, and four pigs. Seven people lived in another underground kitchen, in which there was a little child lying dead. Another apartment contains father, mother, and six children, two of whom were ill with scarlet fever. Where there were " beds," they were simply heaps of rags, shavings, or straw, but for the most part the occupants found their night's rest upon the filthy boards. In many cases matters were made worse by the unhealthy occupations followed by those who dwell in these habitations, especially by match-box making. Even where it is possible to do so the people seldom open their windows ; but were they to do so, it is questionable whether much would be gained, for the external air is scarcely less heavily charged with poison than the atmosphere within. Wretched as these rooms and the common lodging-houses are, they are beyond the means of many who wander about all day, picking up a living as best they

can, and then take refuge at night under a
railway arch, or on a door step. The common
lodging-houses are generally the resort of
thieves and vagabonds of the lowest type, and
some are kept by receivers of stolen property.
But hundreds of the homeless and houseless
cannot even scrape together the wretched twopence required to secure them the privilege of
resting in the sweltering dormitories of these
miserable dwellings "dossing on the rope"
as the slang goes, so they congregate upon
the stairs and landings where it is no un-
common thing, as the policeman on the beat
only too well knows, (who, to his credit be it
said, seldom, as the glimmer of his lantern falls
on the poverty stricken wanderers, has the
heart to pronounce the professional "move on,")
to find six or eight individuals in the early
morning.

One of the saddest results of this overcrowding is the inevitable association of honest
people with criminals. Often the family of a
respectable working man is compelled to take
refuge in a thieves' kitchen, so that their con-
tinual contact with the very worst of those who
come out of our gaols is a matter of necessity.
Wrong doing and vice thus become habitual,

for as habit is to the individual so custom is to
the mass. Then, clearly there is such a thing
as the being held in bondage by mere custom,
and, no doubt, the pauper population are more
or less in this case, so that the great thing is
to change that which has most influence, then
to rouse the energies of the people, and they
will afterwards pull themselves out of the groove
in which they have become torpid. Their
confidence must be won by witnessing the
efficacy of Political Science as seen in its
servants' desire to carry out the intent and
purpose which its principles posess.

In past times every reformer has had to face
the insurmountable barriers of ignorance and
prejudice, and has been subjected to bitter
persecution ; and now, notwithstanding the
" civilization" of the most enlightened age the
world has perhaps ever seen, he has still to face
prejudice, in addition to selfishness and interest! These are the great obstacles with
which a zealous reformer has to cope. There
can be no doubt that there are numbers of
habitual criminals who would never have become such, had they not, by force of circumstances, been packed together in these dark
dens with those who had previously been

hardened in crime. The public-house is not altogether as some social reformers would have us to believe, the sole cause of this state of things in our capital, for this evil existed long before the public-houses were so abundant. But who can wonder at the attraction that the public-house offers to the tired toiler in these parts, bringing him as it does a temporary forgetfulness of his normal misery. Contrast it for one moment with any one of the abodes which we may find in the fetid courts behind, and you will wonder no longer that it is crowded. How can the wretched proletarian be expected to resist so overwhelming a temptation although it gives only a *temporary* oblivion from the constant contemplation of a state of things dreadful beyond comparison?

In such cases who can wonder that immorality, and all kinds of vice and depravity have their schools here? Children who can scarcely walk are taught to steal, and they are mercilessly beaten if they come back from their daily expeditions without money or its worth. Still, it is capable of proof that the honest far outnumber the dishonest, and this seems most extraordinary, when we know that a child seven

B

years old is able to make 10s. 6d. a week by
thieving.   Compare this with what he can earn
by honest work, such as match-box making, for
which 2¼d. is paid for a gross of boxes, the
maker having to find his own fire for drying
them, and his own paste and string.   Before he
can gain as much as the young thief he must
make fifty-six gross of match boxes a week, or
1,296 a day.   That is impossible for even a
grown up man or woman to do, and they can
rarely make more than half that number.   A
woman, who is making tweed trousers, can earn
a shilling a day ; but what does a day mean to
this miserable slave ?   Seventeen hours ! from
five o'clock in the morning to ten o'clock at
night ; without even a pause for meals.   In St.
George's in the East, large numbers of women
and children (some of the latter only seven
years old) are employed in sack making, for
which they get a farthing a-piece.   Others are
employed at shirt finishing at threepence a
dozen, and by the utmost effort can only earn
sixpence a day, out of which they have to find
their own thread.   Other women obtain, at
Covent Garden, in the season, one penny or
twopence a peck for shelling peas, or sixpence
a basket for peeling walnuts, and they do well

if their labour brings them tenpence or a shilling a day, and upon that pittance they have to live. And with the men it is little or no better. " My master," says one man, who was visited by a recent writer in *Fortnightly Review*, " gets a pound for what he gives me 3s. for making." And it is easy to believe this, when we know that for a pair of fishing boots (which will be sold for three guineas,) the workman only receives 5s. 3d. if they are made to order, and 4s. 6d. if made for stock. The same author speaks of an old tailor and his wife employed in making policemen's overcoats. They have to make, finish, and press them, put on the buttons, and find their own thread ; for all this they receive only at the rate of 2s. 10d. on each coat. This old couple work from half-past six in the morning until ten at night, and between them can just manage to make one coat in two days, and keep the wolf from the door, until the merciful angel calls them from the stage of life. This system of apportioning different work to the various departments by the employer is called the sweating system.

There are 2000 sweaters in the East End. The payments to the work people amount to 13s. 3d., leaving 10s. 9d. to the Sweater. Take

again the coster trade, a very large industry in
London, employing many thousands : its districts are Somers Town, Leather Lane, and
Golden Lane ; the cruel monopoly of markets
held in the City deprives London for seven
miles round of the advantages which would be
enjoyed if this, together with the many
other similar unjust imposts, were done away
with.

The rents, which are paid for some of these
rookeries, are very excessive, while honest
labour seldom gets its just reward. The
Reverend Archibald Brown who, together with
his missionaries, has done a noble work in the
East End, tells us that people had doubted
the accuracy of the reports presented by
the missionaries, and that he therefore devoted a considerable time to personal visitation
and enquiry ; when he found that but a small
proportion of the truth had been told. It is
only just to say that the Bell Lane (Whitechapel
District) is being improved by the Board of
Works, but only after great pressure had been
brought to bear upon its members. The Board
were even threatened with exposure unless the
refuse was removed with greater regularity and
the water supply increased. Speaking of water

reminds us forcibly of another instance of unearned increment, as seen in the great monopoly
of the New River and other Companies, which
supplies that necessity of life to great portions
of the Metropolis.[1]

No doubt a more extended system of outdoor relief to the aged and crippled throughout
the United Kingdom, would be an alleviative if
administered with discretion.

But the remedy here prescribed for the removal of these unhealthy dens, in which thousands of otherwise strong able bodied young
men and women brook out an empty existence,
has no connexion or relationship whatever with
our national Poor Law system.

The 25 London casual wards are healthy
enough. They too, however, have no application to the cure.

In fact, the Poor Law system in all its phases

---

[1] This New River Company was divided into 72 shares,
which once upon a time sold for £100 each, but which are now
worth from £9000 to £10,000 a piece or from £648,000 to
£720,000 altogether.

At this present moment the frightful state of the outbuildings
alone, and the water supply in some Districts, would be an out-
rage on civilization were the age merely pagan or even bar-
barian.

Where is the "fellow feeling which makes one won drous
kind" in this age of so called Christian refinement?

occupies a place quite foreign to that question which we are here treating judicially as well as politically.[1]

Now this state of things exists, and is allowed to continue in a country where the Legislature has justly prescribed punishment for ill-treatment, or neglect of a horse or cow, and where it is not even permitted that pigs be under the same roof as cows, so strict on *sanitary* grounds is the municipal law. In sanitation as also in the principles of political knowledge there is a remedy which is practically detailed further on, but which we have not yet put into force. Not only is this condition of things morally and physically bad, but from any point of view it must be endangering the safety of the State. If it affected the health of wealthy individuals occupying neighbouring property in a rural district or the West End it would be stopped by an injunction in the High Court. An eminent legal scientist has held that people

[1] Most jurists would be prepared to admit that to give free meals to School Board children in those seminaries, (whose parents through want of labour are too poor to pay directly) by increasing the Local rates is within the orbit of practical politics. We find that in the East End 13 per cent are insufficiently fed, and in one School Board district the proportion was 32 per cent.

have the same right to their health that one has to the preservation of a good reputation. Our English Law has for years allowed the would-be injured party to apply in Chancery at once, and without first giving notice to the intending publisher to prevent the publication of a libel. To some, this may seem rather strange, but the reason is clear and not far to seek. A man's reputation is really his property. By judicial publicity we can always clear our character, but we cannot always regain our health.

But the children's misery is the most heart-rending of all. From the beginning of their lives they are utterly neglected. Their bodies and rags are alive with vermin. They are subjected to the most cruel treatment, while many of them have never so much as seen a grass field, and often pass the whole day *without a morsel of food.* In one house, there were nine motherless children. The eldest was only fourteen years old, and all lived in one small room. Further on, in a cellar-kitchen were nine little children in a room one scarcely could see across by reason of smoke and dirt. They were entirely without food, and had very little clothing. If there is one feature, in human nature, which shows itself more prominently than another, it is the brave patience with which the poor endure this state of things. The helpful sympathy (always loyal) which they show one towards another is beyond all praise. Assuredly, their loyalty of feeling has been tested and tried thoroughly by these long years of scientific and social neglect. They are as wretched as can well be imagined, and yet, as a whole, they are marvellously contented. It is not exaggerated, for if we were to proceed with the miserable story, deeper and darker would be the picture. During the last dozen or 15 years, through agricultural depression in our rural districts where some of the cottages are no better than they should be, many thousands more have flocked to our large towns, especially London, so that to all appearances the evil will go on increasing, unless we take some steps to prevent it. Nay more—so hard has the lot of the London poor become, that in the year 1887 the number of deaths caused by starvation or accelerated by privation were 32 : This is shewn by the Parliamentary returns.

In 30 London workhouses there is accommodation for nearly 31,000 inmates : In 15 of these workhouses during the winter of 1887 and 1888 the number of paupers exceeded that for which there was certified accommodation. How is this almost unprecedented state of things brought about—the one cry is—" There is no work," and that answer has a universal application throughout the United Kingdom.

On this being made known to the White-chapel Guardians, a Committee was appointed to consider the advisability of establishing rural settlements, evidently with a view to the cultivation of the soil, and thus to prevent unemployed labour gravitating to London with such baneful results. It was proposed to convene the Guardians of other metropolitan unions to consider a scheme. What form this movement may ultimately take and where it will end is not known.

Another fact would appear certain! Canada will not much longer take our pauper population.

Of Statistics we have had enough and to spare. Yet the gravity of this national question is so great that we ask the indulgence of the reader once more!

Mr. Chas. Booth lately read a paper before the Statistical Society upon the condition and occupations of the inhabitants of East London and Hackney. From the figures given by Mr. Booth, and if we may assume the London popu-

lation within the limits covered by the statistics
submitted to be 4 millions, it is easy to compute its composition approximately.   About
50,000 must be loafers, 300,000 pick up
casual and scanty earnings, 250,000 are receiving irregular wages, 400,000 are in regular
work, as the *Daily News* alleges, for insufficient pay; and continuing says, "We may pretty
safely draw this conclusion that ¼ of London
population is, if not in poverty or want, at any
rate below the line of plenty."   As we have
the vote, we are, at least, in some sort of way,
the servants of justice fighting under the Royal
banner of this great cause as before described.
These evils lie at the door of the people now!
surely it is the last watch of the night!   In
the face of an evil so great it is impossible for
the better institutions of this country to do their
work effectually, [for if a thing is worth doing
at all it is worth doing well].   The poor must
live somewhere, and they must inhabit where
the great centres of industry are to be found, or
in a locality where there is a cheap and ready
access to them.   What of our surroundings can
influence our lives more than our dwellings?

CHAPTER IV.

What has Legislation done?

ENGLISH legislation has in its highest sense laboured to make a life on earth somewhat preparatory to that which must sooner or later follow in a never ending eternity. But in trying to do this, it has been most reluctant to interfere with the liberty of the subject, and the freedom of contract as between man and man. This interference becomes more necessary as population increases. Now Legislation and Administration have but one duty in the vast space of political science, they are so closely allied to politics as to have but one place in the history of National progress. Their function ostensibly has ever been to command what is right in a State, and prohibit that which is wrong, and *secondly* to make and regulate the tribunals by prescribing from time to time fresh rules of procedure ; in short, to regulate society and to maintain it in systematic order. In their endeavours to carry such commands out, and with this object ever in view they have worked on the basis of a few maxims. But mere maxims or abstract principles cannot be quoted in support of the remedy here detailed for treating this great national question. It is contended that when labour is starved we are not compelled to go "from precedent to precedent." However, let us see what statutory remedy legislation has endeavoured to apply. In the reign of James I. an Act was passed forbidding people from coming to live permanently in London ; this is one of the few instances in which, in a time of peace, our written law has wrongfully attempted to interfere with the liberty of the subject within the confines of our own shores. This Act was succeeded after a long interval by three Acts, called the Labouring Classes' Lodging-Houses Acts, of 1851, 1866, 1867, and 1885, the main object of which was to empower the local authorities to acquire land and erect buildings called "Common Lodging-Houses," and they establish a proper and elaborate system of registration of the many common lodging-houses before referred to. These Acts also provided for certain sanitary improvements. Their author was the late much beloved and respected Lord Shaftesbury's father, who, through his long and useful life, toiled to make a wider place for justice in the pages of Political history. Born in the drawing-room of ease, in the springtime of manhood, he forsook the fields of science and literature, although they had already promised fame and success to a great mind, to befriend those, some of whom through no fault of their own, were alone and helpless in life's rough way. Those Acts were followed by the five Artisans' Dwellings Acts 1868, 1875, 1879, 1882, and 1885, which were great factors in the national movement for improved sanitation, and their object is much furthered by the Public Health Act, 1875, the Public Works Loans Act of 1879, and the Sanitary Acts, 1866 and 1874. The Acts of 1868 and 1885 may be regarded as models of legislation, and must have taxed the unflagging energy of their draftsman and author, Mr. McCullagh Torrens, whose name will always be held in esteem for his untiring zeal when the effacing fingers of the legislature shall have ceased to do their work. In addition to these there are the Metropolitan Building Acts, 1855, 1860, 1861, and many others.

The worst of the matter is that such kind of

city property pays the best. It has been ascertained from the tenants that rents are often in arrear, from ten to forty, and in some cases even sixty per cent. are realized upon this class of property. And it is remarkable how very reasonable the local rates and taxes are ; the rates and taxes in the metropolitan parishes as a whole, average 4s. 5¾d. in the pound, whilst in the City proper the average is 7s. 4d. in the pound, being in both cases far below that of most provincial towns. The Disused Burials Ground Act has lately been passed, notwithstanding the great popular desire for acquiring open spaces. As Lord Meath in his " Social Arrows " has said, "why should not disused burial grounds be made resting places for the living ? "

No wonder that the Open Spaces Association and Self-help Emigration Societies and an International State-aided Emigration Association, and the London Sanitary Protection Association should be necessary, but these bring us down to the present time.

What has been done ? About 40,000 people have been provided with houses under these Artisans' Dwellings Acts of 1868 to 1885 inclusive, but this is comparatively nothing.

How is this?

In the first place the great vested interests of the members of the Metropolitan Board of Works, and also the London Corporation or Commissioners of Sewers, as they are called, have ever been the great obstacle; the want of unity among the members of each body, and their apathy and indifference to responsibility, the discarding of the legal and moral obligation of duty which rested upon their individual shoulders, have caused, in a very great measure, things to be what they now are.

The Corporation has said, "It is for the Metropolitan Board of works to do this," and the Metropolitan Board has retorted, "Oh no! it is for you to do it." There has been a want of motive power in each body, and obstinacy and selfishness have been the great causes of it. To any person who has time to carefully peruse the Acts of parliament above referred to, commonly called the Torrens and Cross Acts, it must be obvious that at one time it was hardly necessary to give additional powers to either of these authorities, or to the Local Government Board, had their members been true to first principles; nor need the legislature now tax its ingenuity to establish a precedent of principle with the view to the removal of this great evil, for it yet lies at the door of that science administratively.[1]

This anomaly is the fault of administration ; the onus of its responsibility now rests with those who enjoy a parliamentary franchise, more especially since the Elementary Education Acts have worked their way.

Lord Cranbrook in the House of Lords, in the beginning of the session of 1884, said that the State had duties to discharge in reference to the moral well-being of the people, which duties had been increased since Parliament had made education compulsory. Is not this true ? Are not fit dwellings on a par with education ?

[1] On account of political charges and upheavings on the Continent of Europe, more particularly the expulsion of the Polish Jews, the Whitechapel (Bell Lane) District has been flocked into by those who have always sought shelter and protection in this, our land of liberty, and who, on account of their cleanliness, thrift and providence would be a great boon, had they not settled in a district sadly overcrowded already ; but no patriot can admit that the land is overpeopled !

We have all an interest in each other, and we live out of one another's labour, so that our maxim must be, " Each for all and for himself as part of all," and not " Every man for himself and devil take the hindmost." As we allow voluntary emigration to all, we cannot prohibit the influx of foreign paupers, either Polish Jews or any other nationality. We are ourselves composed of five, if not more nationalities ! Should the number increase, or Canada close her doors against us, Lord Meath's suggested Poll Tax would be a just remedy.

Surely they ought to be preferred to it as infinitely more affecting the tone of society and the moral well-being of the people.

With befitting integrity, Mr. Torrens said at the Mansion House council, in January 1884, that if they wanted to build cottages in an over-crowded town they must pay an extravagant sum for the sites of the wretched houses which they had pulled down, and that this had prevented builders coming forward to replace those which were destroyed. Now where the original lease had expired it would be well to allow the space to be purchased at a fair valuation, or by arbitration, by the London Corporation, or by the London County Council, and such empty space to be planted in a similar way to New Square and Lincoln's Inn Fields, but always to remain open, and unbuilt upon. And in many cases where, on account of the price asked, being thought too much, and the spaces unable to be purchased for planting, after the slums had been removed, under the Acts, to which reference has before been made; the empty spaces themselves, although unsightly, would be healthier and preferable to their being again filled up with ungainly buildings; in such cases the open

spaces ought justly to be allowed to stand vacant
and be as a monument of individual greed and
selfishness.　This would certainly increase the
rates and taxes, but not unduly so, for comparatively speaking, London as a whole is not over
burdened with taxation, and would it not increase
the health of the people, and gain time?　Mr.
Torrens has ingeniously raised the question of
taxing the ground rents to aid in the carrying
out of these necessary changes.　This can be
legally done, for the houses (not the land) pay
the increased rates and taxes out of the rents,
yet the land rises considerably in rental, but
escapes its just proportion of local burdens.
Both land and houses serve one purpose, and
surely the branch savours of the root, but the
mortgagee can never in principle be saddled
with local taxation.[1]　Apart from the fact that
this valuable building property is owned by a
few private individuals (a few being members
of the legal profession), and upon whose interests
we have no right to trespass by any scheme of
confiscation; it also in a large degree belongs to
a few great corporations, such as the Ecclesi-

[1] Although it would be difficult, yet on the same principle and
with equal justice ought the recipients of fines (which are
capitalized ground rent) to be made to bear their due share of
the burden.

astical Commissioners and the Governors of such institutions as the Foundling Hospital. Now in a legal point of view these parties, being trustees for charitable purposes, are often difficult to approach. They can not incur any responsibility in such cases ; but would it not be as well for them sometimes, more especially where the leases have expired, to plant a portion, and thereby enhance the value of their adjoining property. Why should not the Ecclesiastical Commissioners obtain an Act and lead the way ? It is well known that they own some of the worst of these slums !

London Local Self Government, before Mr. Ritchie's Act.

LET us now see what system of local self government has been adopted and pursued where
these troubles exist. This has been taken from Mr.
Sydney Buxton's Handbook to Political questions.
When the last census was taken in 1881, the metropolis contained 3,813,000 souls, its rateable annual
value amounted to £28,000,000, towards which the City
proper, with 51,400 inhabitants, contributed £3,590,000,
leaving the annual sum of £24,410,000 to be raised
outside the City proper.

Now the authorities, who, between them, controlled
and governed the metropolis, were as follows:

The Corporation of the City of London consisted
of 26 Aldermen and 206 common Councilmen, which
had full municipal authority over the City, and levied
therein rates and taxes.

II. The Metropolitan Board of Works as consti-
tuted in 1855, forty-three of whose members were
elected by the vestries, of which there were twenty-
three, and three by the common council. It,
throughout the extra-city metropolitan area, controlled
the main drainage and sewerage, the Thames em-

bankment, the bridges, street improvements, building, naming and numbering of streets, and looked after dangerous structures, artisans' dwellings, commons, parks, and open spaces, the several fire brigades, nuisances, explosives, the conveyance of cattle, infectious disease, floods, and other similar casualities.

III. The twenty-three vestries and fifteen District Boards, the members of which, numbering 2,446, were elected by the ratepayers, a third of them retiring each year, who controlled the paving, lighting, watering, branch drainage, cleansing and sanitary matters, in their respective districts, and assess the houses, and levy parochial rates for these purposes, and for poor law administration, as well as meet the precepts issued by the Metropolitan Board of Works and the School Board ; the rates varying in different parishes from as little as 3s. 4d. to as much as 7s. 4d

IV. The School Board, which has charge of the elementary education of London and whose fifty members are directly and publicly elected by the ratepayers every three years.

V. There were thirty Boards of Guardians elected or appointed in different ways who had charge of the poor law administration.

VI. There was also an Asylums Board which consisted partly of Guardians, and partly of Nominees of the Local Government Board to look after the sick poor.

VII. There was the Thames Conservancy Board, non-representative, and which had control of the River Thames.

VIII.   In addition, the Home Secretary had control over the Cabs, and the Police force outside the
City; whilst within the City they were under the
control of the Corporation.

IX.   The Water and Gas Companies which, however, are private concerns.

Now, even to the casual observer, this medley of
systems could not appear methodical, and on a closer
examination we shall discover that it was not so, for
entire want of method prevailed.

The happening of fires was good instances of the
confusion of responsible authorities.   The fire brigade
was under the control of the Board of Works, the
Police obeyed the Home Office, the Salvage Corps
was under the command of the Fire Insurance
Offices, the Turncocks were the servants of the
Water Companies, and the thoroughfares were the
property of the Vestries.   Again, the lighting of
some of the parks was in the hands of the Commissioners of Woods and Forests, whilst others were
lighted by the Board of Works, and the Board of
Trade had supervision over the Gas supply.[1]

[1] See Appendix.

CHAPTER V

The Remedy.

IF the Disused Burial Grounds Act were so altered as to prohibit building, without the Vestry making compensation to the Owner, those places might then be planted with evergreens or trees. A measure ought then to be passed, appointing a select committee with full power to act, who should appoint architects and engineers, as inspectors, to co-operate with local medical officers of health, in selecting sites for the erection of proper buildings, upon plans approved by such select parliamentary committee. The land purchased for this purpose ought to be free from ground rent and freehold of tenure and inheritance, and should be conveyed to the First Lord of the Treasury, and paid for by the State out of the Imperial Exchequer. It would be advantageous for such lands to be situate a few miles from the various centres of industry if (as might be) a little beyond the bus, or railway terminii, yet within such a radius as to give a quick access to the city in these days of Railway and Steam Engines.[1] These houses might some times be unoccupied, but that " near London " would never be for long; they might and perhaps ought to be erected in the rural districts, and, as for giving compensation to the tenants on eviction, their better houses should be deemed a sufficient compensation, for it would only be another instance of interfering with the liberty of the subject for his own and for the public advantage. Now there is no comparison between the price paid for labour and the high rents given for these slums. A former Government by acquiring shares in the Suez Canal brought the loaf and food generally pretty much on a par with labour, whilst clothing has never been so cheap, and the facilities for an economical yet substantial education under the London School Board curriculum, point to the fact that the other essentials to human existence, namely, adequate dwellings at reasonable

[1] There are some who will say that the railway fares would be beyond the means of those who dwell in these sinks of pestilence, then the Government must exercise a prerogative long since adopted and compel the Railway Companies to reduce their 3rd. class passenger rates, but it is not likely that the pressure of Local competition and the prospect of an increase of passenger traffic would be a sufficient incentive to the Companies to do so without parliamentary interference,

rents in healthy metropolitan districts, have not yet been had, and which when supplied would be the removal of the anomaly now pointed to.

Eviction, no doubt, is sometimes a difficult problem; an order from a Magistrate, on the affidavit of the local Medical Officer of Health, or one of the above-named Government Inspectors as to the unhealthy and uninhabitable state of any rookery, might be made sufficient for this purpose. It will be seen that the whole initiative as to demolition of each of these rookeries would be taken by the Government, through the agency of this committee, three months' notice being given by it to the owners, who would, on their tenants being removed to their new habitations, be paid and compensated by the whole municipal authority. The removal of the slum would then follow, and it might be planted or left vacant as the voice of the majority of the London County Council should determine.

This course would more equally divide the burden, for as a Government is the guardian of the people whom it controls, is it not clear that the lowering of the before-mentioned death rate and infant mortality ought to be one of its first duties? In order to effect this, not only

must the State make provision for those means
of physical training by the use of public gym-
nasiums and baths which are admitted to be
equally as necessary for the metropolis as
elsewhere, but sufficient habitable dwellings
for her people ought first to be built. The
slums having been removed under the provision of the Cross and Torrens Acts, the
old sites would not all again be built upon,
so that there would require to be the new
Act passed to give the Government power
(which in theory it has long had) to build the
new houses in a rural or semi-rural district.

It may be said that such a proceeding as
this, taken by the Government, would be
without a precedent in the history of this
country. Whether or not the circumstances
warrant it, must be left to the better judgment
of the people, but the suggested remedy has
been adopted elsewhere, and is not altogether
without a precedent here.

Popular Contract with the Legislature.

Written expressly for Working Men.

" THE original and continued existence of all
" society," as Stevens tells us, " has an implied
" contract as its common basis. The relationship be-
" tween the Government and the people rests on a con-
" tract, mutual and reciprocal, which involves the legal
" and moral obligation of protection on the one hand,
" and reasonable subjection on the other. The Legisla-
" tive power presumably promises three things as an in-
"ducement to the people to enter into this arrangement :

> " (1st)   To govern according to law,
> " (2nd)   To execute judgment in mercy,
> " (3rd)   To maintain a true religion.

" The compact is, "that so long as the Government
" affords protection, so long will the people demean
" themselves faithfully." " The original founders of
" our Government might have provided for an elective
" monarchy ; but they chose, for good reasons, to have
" a succession by inheritance. It must be owned, the
" same writer reminds us, that an elective monarchy
" was the most obvious, and best suited to the rational
" principles of government and the freedom of human
" nature, and, accordingly, we find from history, that, in

" the infancy of almost every State, the leader or prince
" has usually been elective. And if the individuals
" who compose a State could always remain true to
" first principles, it might be plausibly argued that
" elective succession were as much to be desired in a
" kingdom as in any other inferior community ;
" the best and wisest man would then be sure of re-
" ceiving that crown which his endowments had
" merited, and the sense of an unbiased majority
" would be dutifully acquiesced in by the few who
" were of different opinions. History and observation,
" however, inform us that elections of every kind (in the
" present state of human nature) are too frequently
" brought about by influence, partiality, and artifice,
" and even, where the case is otherwise, these practices
" will be often suspected, and constantly charged upon
" the successful by a splenetic disappointed minority."

Is not this an instance in which English law has
availed itself of the weak features in our nature to our
predisposition to mere custom for acquiescence? And
yet it allows a state of things to exist in our midst,
which is so highly productive of discontent and
socialism until there comes the crisis which has so
often overtaken other continental nationalities.

Then, again, those who occupy the higher official
posts in life are the objects of envy to others in the
lower stations, and this is one reason why the House
of Lords is not elective, and is again an additional
and forcible instance of how we English have steered
clear of appealing to, or rather trusting ourselves to
the frail instincts of human nature. But so long as

the Government allow this dreadful state of things to continue under the cover of thickly populated houses, so long will London slums naturally ever be the birthplace of crime, socialism, discontent and revolution. Surely recent events and our newspaper reports conclusively prove this. Under these conditions, can this popular contract be observed with that integrity which is otherwise such a marked feature in the principles of political science ; a science which has not only in theory so long maintained the completeness of its principles, but has claimed effectually to carry them out by impartial administration in our Courts of Justice ? Do we not lay ourselves open to the worst and most pernicious influences by allowing these hovels to continue ? Ought we not rather to endeavour to make the conditions of society such as to prevent one portion getting out of joint with another ? But does not this twofold compact upon which English legislation is based rest upon that confidence, that faith and trust which its subjects ought to have in each other, as also in the competency of its servants to rule society ? And what is this mutual and reciprocal conviction of confidence but love ? it is that feeling which we ought all to have one for another if we lived true to cardinal principles ; and is not love the beginning of religion ? The value of this confidence, as it ought to exist between the individuals who compose Society, or between a Government and its people, can never be told, only by those who have once enjoyed it and afterwards (through no fault of their own) experienced its want in private life equally as in the vicissitudes of political strife.

It is the life blood and foundation of systematic legislation! It secures the peaceful succession of events from one generation to another! It was the pride of by-gone Politicians, and it is the hope of those who are now!

It is the reverse of injustice, pride, and arrogancy, and it cannot exist with them.

Now this feeling of confidence cannot exist along with anarchy and confusion which are the result of socialism and popular discontent, brought about by this overcrowding of the masses in the Metropolis and other cities and towns, assuredly the existence of these rcokeries must be the anomaly of political science.

In the reign of Queen Elizabeth, in order to en-courage native industry in Lancashire and the West Riding of Yorkshire, an Act was passed prohibiting the exportation of wool in its raw state, so as to give employment to labour, but the great increase of foreign importation of that commodity, during the last sixty years, had rendered the effect of this statute inoperative. By another Act in the reign of Charles II., it was ordered that the dead should be dressed in woollen clothing before burial. These were both interferences with the liberty of the subject, on the grounds of public policy and political ex-pediency, to encourage the staple trade of a particular locality; which they successfully effected, for the result has been that these two together gave content-ment, independence and prosperity to that district,

If a Government have the power to preserve society at home, by the maintenance of a large body of Police, a Navy, and Soldiery, it is its property, for the reasons before given, to build adequate dwellings in healthy districts and let them at reasonable rents. What was the object? and what has been the result of the purchase of the Suez Canal shares by the British Government? The loaf has been brought to our doors at prices unprecedented in the annals of commerce; has not the Legislature anticipated our wants, and taken time by the forelock in that great undertaking? The Government has interfered to cut down railway charges and rates on a similar ground of public policy. It has bought the telegraphic system, and might exercise its power to buy up the railways and work them, and has constructed railways abroad, as for example at Quetta; all on the like ground of public policy. Why, for the same reason, can it not build suitable houses? for if a Government has authority to purchase a portion of a canal for the national welfare, it has surely power to make the conditions of every day life such that London poor can perform their part of that contract to which reference has already been made. The neglect to demolish these slums is doubtless a failure in the performance by the Government of its part of the mutual and reciprocal compact before referred to! Burglaries are not decreasing in London! Petitions in the forms of Bills to the Sovereign direct, have fallen into disuse—they are now presented to parliament! If her Majesty as " parens patriæ " had been applied to in person, this great anomaly might have been removed long ago, indeed it is more than probable that it would have been !

Legislation has always claimed to be consistent and methodical in its principles as applied to society, but surely it has failed in the one instance before us. What has not the Government, who grants large sums for educational and religious purposes, power to do on this one ground of public policy ? Does not our constitutional form of Government show that continuity is a main object of legislation, yet it allows the body politic upon which it *operates to degenerate* in its very centre, and permits charitable and philanthropic schemes of emigration to rob the mother country of its backbone. Technical education for trade purposes on account of foreign competition and quick steam transit will avail little more than enable us to keep our head above water.

CHAPTER VI.

𝔓ractical 𝔄spect.

THERE are who will say that thrift and providence ought to have made the Metropolitan children of toil equally as independent as they have made those of Lancashire and the West Riding of Yorkshire. The answer comes that no staple industries, such as the Cotton, Woollen, and Carpet manufactures have ever given them the chance. The various Building Society Acts (the first of which was cradled in Rochdale), the Co-operative Stores, and the Industrial Societies Acts have never reached to help them, as they have done the men and women of Lancashire, to comparative comfort and ease. The great Lancashire manufacture of machinery, engines, and boilers, with the consequent demand for honest labour, has never extended its beneficent hand to London. The period of Lancashire's prosperity, from 1820 to 1870, was unshared for the most part by the inhabitants of these fearful dwellings, the mere creatures of chance. The great prosper-

D

ity of insurance companies is a sure sign of thrift ;
they are characteristic of the times, but they have
not benefited these children of toil.   The members of the medical profession of the last generation ignorantly bled fever patients to death :
they were taking from them the very means of
life sustenance ; if they were to do so now they
would be indicted on a capital charge, and suffer
the extreme penalty.   But is not the Government nearly in a similar way responsible for
allowing a state of things to continue, which produces this incomparable infant mortality and degeneracy to London people whose nationality
has carried Christianity to the heathen, and education and prosperity to the declining nations of
the East ?   This infant mortality and degeneracy must be the wonder of the inhabitants of
Lancashire and Yorkshire (West Riding) where
very many working men and women own the
house they inhabit, many indeed being proprietors
of from 6 to 8 model cottages a piece, which they
have acquired through the aid of some local
Building Society (or their individual frugality
and thrift), and where the Post Office Savings
Returns show the safety of the nation in
those parts.   What a contrast ! Can the spread
of education be otherwise than productive

of an increasing desire for freedom? Can it
be so in London? These facts are the
State's sheet anchor; why should the Metropolis be the exception? "Why should London
wait?" Is not the Legislature itself a slave to
habit, and held in bondage by custom? Has
it anticipated the turn of events in the destiny of
man, or taken time by the forelock in this great
question? Now the duties which are performed
by a local authority are next in importance to
those observed by a Government, and a local
authority has power to build model houses and
furnish them, as can also a corporation with the
Treasury's consent. The province of our national Poor Law system was never really intended to extend to young and able bodied men
and women, who are quite able to earn their
daily bread by the honest sweat of their brows,
and who would successfully do so but for
cruel trade monopolies and the glut in the
labour market in our conglomerated Metropolis. The workhouse therefore is no proper
place for them. The Church has not only
educated, but also at one time systematically
attended to the daily necessities of the poor.
But this question in its present dimensions
in London is utterly beyond the reach of the

Church, and the scope of the best organised Charity.

The German Government, in order to provide against poverty, have adopted a system of compulsory individual insurance. It is, however, beyond the present spirit and authority of English legislation to impose such a burden upon her subjects. Poverty has never been a "crime" in the eye of English law. Shortly after the passing of one of Viscount Cross' Acts, a circular was issued asking the different authorities to give the medical officers directions to make an official return to the Metropolitan Board of Works as to all areas which they considered unhealthy, with the view that the Metropolitan Board should act upon them. These circulars stated that the Secretary of State desired to be furnished with a copy of these returns, in order to be able to lay them before Parliament. Well, the answers came in very slowly, and application after application had to be made before any reply could be got at all in many cases, and none ever came from the Westminster district : and in the case of Bermondsey, it was five years before any reply was given.

There is no body of men, who in such a large way, yet at so small a pecuniary return to them-

selves, have benefited society more than the
builder and contractor. How few (if any) have
made a competency out of their calling. The
Leasehold Enfranchisement Bill of Lord R.
Churchill and Mr. Broadhurst is timely and well
calculated to prevent jerry building for the future. The Canal Board Acts is a twin sister in
legislation to the late Lord Shaftesbury's Working Classes Lodging-Houses Acts ; whilst Mr.
Plimsoll by his Shipping Act, has in some way
secured the safety of the lives of those who go
down to the sea in ships ; but the security of the
lives of those who live on the firm earth has not
yet been insured by that science which, from the
earliest times, has claimed to regulate society,
and to give equal rights to all men. All honour
to Mr. Samuel Plimsoll for his great achieve-
ments in maritime affairs, with which this subject stands very much on a par. When one
thinks and is led to look more closely into this
great question, the frailty of the human animal
is obvious as well as his great love of self, especially his improvidence when compared with the
foresight of the ant ; more particularly so when
it is contrasted with the long continued neglig-
ent indifference of our Parliaments year after
year.

"Reasoning at every step man treads
    He yet mistakes his way,
    While meaner things by instinct led
    Are seldom known to stray."

Is it not a truism that law and morality are
the same thing? The discharge of the legal
and moral duty, which we owe one to another,
is nothing less than the observance of God's
command, and is else but a part of the duty
which man owes to his Creator. Perhaps some
will say that as a people we are not religious
and charitable : if we are not, there never has
been a country that was. Have we not the
foreign and home missions of various Christian
denominations? Have we not had the Bulgarian
relief fund, the Chinese relief fund, the Indian
relief fund, all in times of famine in those
countries, but of plenty at home? Then why
has this great question been neglected? What
is everyone's duty is no one's, but is there not a
continuous famine in the Capital and large cities
of our own country? Does not the fact that
the London Unemployed relief fund amounted
to the enormous sum of £65,000 in aid of
20,000 Unemployed show the destitution which
exists? And the many free breakfasts prove
that the labour market in London is over

supplied. We have, as a nation, at all times allowed real property to be a source of investment, equally with consols and the Post Office savings: and the Government cannot now interfere to regulate its value (excepting on urgent grounds of public policy) without acting in bad faith and committing a breach of that contract upon which English society rests. True, it is a legal principle that "private right must be subservient to public utility," that is, whenever and wherever property is or has been required for public purposes : but in these cases there has always been an adequate value paid, in addition to a proper compensation (usually fifteen per cent., sometimes more) for enforcing a compulsory sale. The land is therefore for the people, subject to the just rights of those individuals, who, relying upon the strength of this mutual and reciprocal arrangement, may have invested their hard-earned savings in the purchase of either a humble cottage home, the shares of a building society, or the broad acres of an agricultural property. It has been said that the land laws are responsible for this anomalous state of things in London, and the following extract from Mr. Williams, the great writer on real-property-law, would to the unwary appear to favour this view, and it has been lately so used in a leading periodical.

" The first thing the student has to do, is to " get rid of the idea of absolute ownership. Such "an idea is quite unknown in English law. No " man in the law is the absolute owner of his "lands. He can only hold an estate in them." This is doubtless so in theory, but it never has interfered in practice with the utilization and enjoyment of property. Had the Vestries and other authorities been true to first principles of right, this great evil would never have existed in its present stupendous magnitude. Great writers show that legislation, as a christianizing power, has claimed to give a remedy for every breach of right and to regulate human affairs. It has surely neglected London, where its legislative-tribunals have long sat at Westminster, and where ninety per cent. of its Devotees (Counsel), learned in this great science, have worked to lubricate its wheels to make it keep pace with the growing wants of an ever increasing society, and where its judicial tribunals have co-operated to make men just, one towards another. It is there where this great evil exists in the very heart of the nation.[1]

[1] The Lord Mayor is allowed by the City Corporation

The mental and physical degeneracy we so greatly deplore can never be checked so long as the rural population flocks to these unhealthy dwellings. " Mens sana in corpore sano " is an axiom which must not be forgotten. If its neglect continues, a speedier degeneracy will overtake us.

£10,000 a year out of which the expenses of the Show (wholly unnecessary) are defrayed.

The City Corporation and its rich Livery Companies or Guilds have a yearly income of a million sterling. From a report of a recent Royal Commission half of that amount would appear to be applied to private purposes instead of being expended on such like undertakings as Public Baths, so says the *Daily News*. We think habitable dwellings ought to be supplied first.

CHAPTER VII.

Sketch of Progressive Legislation.

LEGISLATION is a deep research, and foresight always had a prominent place in its history of progress, but action by anticipation has been its chief aim in assisting us to provide for our national wants and necessities. " Providence helps those who help themselves." When Geo. III. was but newly crowned, men and women were sold as slaves by advertisement in Liverpool. The liberty of the press has been strengthened yet prescribed ; comparative freedom given to speech and debate ; the Corn Laws have been repealed. The Ten Hours Factory Act for which the late Lord Shaftesbury's father and Mr. Fielden fought, these have been passed ; and when taken together with the Common Classes Lodging-Houses Acts, and the hitherto inoperative Artisans' Dwellings Acts, the Canal Boats Act, and Mr. Plimsoll's Shipping Acts, make us conclude that clearly there are strong Christian traits in legislation as a great civilising power.

Nay, more than this ; it has long since made the judges and magistrates completely independent of the sovereign power ; but, instead of having State surveillance over these slums, it has left the medical officers of health, by reason of local influence, in such a position that it must have required more than Roman courage for them to have done their duty.   Perhaps they may have done their duty in the majority of cases. Bermondsey, which is one of the worst districts within the area of London, with a population of 86,000, has only one local sanitary inspector. The Legislature has by the Reform Acts of 1832 and 1867 permanently abated the influence immemorially exercised by wealth and power in local tribunals, as also in the councils of the State ; and yet it allows the vested interests of London property owners, when sitting in authority in the vestries, to oppress the weaker members of the community, and with " might," in such a subtle yet prominent form to override justice and equity.   It is this municipal power of selfish individuals in the vestries which has caused such a want of motive power.   During the last twenty years it has been the aim of English legislation to raise woman to a level higher than she has

known before in the whole history of the great
nations of Asia and Europe. It has been and
still is its noble intention to give the greatest
possible independence to woman by protecting
her and placing her above the unbridled aims
of a selfish world. The Married Woman's
Property Acts, the Elementary Education Acts
and the Criminal Law Amendment Act show
this; although wealth or even the smallest
dowry might not be her portion in life, still the
intention is apparent to give her an independence by even protecting her daily earnings.
All this, however, is only in keeping with the
forward march of civilization on the broad lines
of which an advancing Christianity is travel-
ling. Enough has been said to show that
modern Legislation is influenced by ideas of
natural justice as it heeds the principles of duty
and conscience, the moral law of God; or it
could never have steered English society for
one thousand years over the many shoals and
quicksands of destruction which ofttimes threatened to envelope it in the darker ages of ignorance and superstition. This being the state of
the East End, what good can the School
Board do to these children of toil, who
have neither physical strength nor mental

vigour to enable them to keep pace with the times?[1] Surely Legislation is defeating the very end and purpose for which it has ever claimed to labour and fight.

Legislative science is, to some extent, what we make it, and we are what we make ourselves. It can never be said to have fulfilled its mission until this great blot has been removed from the pages of its history. An eminent professor of equity jurisprudence at one of our Universities has said that we want a philosophy of law. Surely not, until this evil has been first removed. And although the year of Her Majesty's jubilee has passed, it cannot but be hoped by all patriots, that the closing quarter of the nineteenth century may not be permitted to hasten its way to an end leaving this cankering mass still in our midst.

The question may be asked, are we our brother's keeper in the eye of the law?

Yes! certainly we are our brother's keeper, and

---

[1] It is our town population, who, the Essayist, at our recent Scientific Congress, told us, are worn out at the third generation. From whence is this deficiency supplied but from the rural districts where Agriculture is the chief employment. Society in all ages has been recruited, and the various departments of national life have been replenished from behind the plough.

this responsibility the Public Health Act, Vacci-
nation Law, the Factory and the Elementary
Education Acts are recognising to a marked
degree, and the world is growing better.

There are some who will venture to say that
primogeniture and the laws of entail are the
cause of this state of things. This also is a
delusion, for property which has come to its
owner through either of these channels, can now
be almost as easily sold as a pound of cheese
can be purchased at a grocer's shop, and this is
so even if the owner be a minor. Whatever
great writers and politicians can say against
primogeniture and the laws of entail, neither
operate in the slightest degree against the
principle that " private right must be subservient
to public utility," nor prevent the constant
transfer of ownership of property as such
changes become necessary or expedient. The
former (that is primogeniture) is commendable
on account of its great certainty, and when
taken together with the statutes of distribution
of personal property, has worked no injury to
society whatever, but has prevented many
family quarrels. The latter (entail) is simply the
exercise of the right which we all have, namely,
to do as we wish with our own, extending to

poor and rich alike, and through the medium of a trustee it can be applied to money as well as to land and buildings. A father has a perfect right to give only a life interest to a drunken or spend-thrift son, or similarly to protect his portioned daughter and her family against the menaces of a husband of like character, so long as he does not block the progress of society. On the other hand it is only our love of greed, man's sordid motives of gain and pleasure which do injury alike to ourselves and society. A forcible instance of this greed occurred lately. A land owner refused to sell his land situated at the entrance to a town until by the equitable and fair-dealing ways of the neighbouring owners (who, acting upon their architect's advice, sold at from 2d. to 10d. a yard, according to eligibility of situation, for cottages and mills on perpetual ground rents), his estate became surrounded, and the value of his seven fields was beginning to rise quickly, more especially as trade soon afterwards became brisk. When the family were asked why they had not taken advantage of the demand, they answered, " That is the result of the laws of entail." Most individuals in such a case of boycotting would have advised the builder and contractor to make no application, but leave the seven fields open so as to afford additional light and air to the houses, which had previously been erected on the closes of the open and fair-dealing owners, until the principle " that private right ought to be subservient to public utility," was forced upon this avaricious gentleman, who was, however, it must be admitted, a great exception to the generality. Now when this gentleman was applied to, scarcely ever could a definite answer be got from him, until, after repeatedly waiting upon him, the inquirer grew tired, and ultimately succeeded elsewhere. Afterwards by degrees his seven fields became surrounded. No doubt inwardly he thought that success was beaming upon him her sweetest smile. In the meantime however, whilst his seven fields have brought him a very inadequate return for their value, the public mind has been educated on some political questions, in a manner which may vitally affect his future interests. Had a gas or water company, a corporation, or a railway company applied, they would have served him with a notice to treat, and take possession, under their statutory powers of compulsory purchase con-ferred by the Lands Clauses Consolidation Acts. Not many months ago there came to the knowledge of a clergyman a case which really leaves this far in the background. An official employed by a wealthy, commercial, but sadly overcrowded corporation owned some property in its centre, which he apparently believed the authorities would shortly require. A few feigned or collusive deeds were made between himself and a friend, the price (which of course never passed between them) being considerably raised on each transaction. Now when applied to for his property the official would with apparent generosity be able to say, you can have the property at the price it has cost me. Which was no doubt the object of the counterfeit transactions. [1]   These individuals would destroy

[1] For is it not the inflated price which would be asked by this official which causes the local authority to be so diffident in seeking to acquire possession of his fetid tenements wherein dwells the seething mass who too would have the price to pay, in the shape of an increased annual municipal burden? Now the rates and taxes are often highest where these unhealthy dwellings are worst. Supposing, therefore, that this official's property is situated in London, the absence of our natural sense of justice towards these unfortunates is seen in our allowing them to occupy them, and inflict an infant mortality and degeneracy upon humanity—nay more—the vacant slums themselves are unhealthy to the neighbourhood.

Is it not therefore clear that the official becomes rich out of the vitals of the helpless, the fatherless, and the widow? He is deaf to the voice of conscience ! heedless of the weight of moral persuasion ! This case is as bad, if not worse, than that of the states and tyrannise over Society to gain their own ends.

Now, it is well known that, as a class, the English agricultural land owners are just and fair, and should there be a few exceptions, we have, so far as farming land is concerned, the Agricultural Holdings Act to apply to. Some time ago there came to the knowledge of the writer the case of an agricultural land-owner, who, as lessor, for some years had received an excessive rent which had always been fully paid up by the tenant. On the expiration of the term the landlord proceeded to have the farm valued for the past years during which prices had been low, when it was found that £1,500 had been paid over and above what could possibly have been made, whereupon a cheque for this amount was forwarded by the proprietor to his late tenant. This is one of the many instances which could be furnished where the owner has practically carried out the principle

Levite who came to look on but who ultimately passed by on the other side !

It is a clear case of the false weight ! the uneven balance ! ill-gotten gains ! and such actions cannot but have their reward in the administration of a retributive justice hereafter. Why should not English Law administer that retributive justice, as we claim it to be a Christian Science ?

of duty and conscience. And bearing in mind that, as there is not, nor ever has been, a law compelling agricultural landlords to repair and maintain the buildings, as between themselves and their tenants, and as many such proprietors have voluntarily spent large sums in so doing, in return for which they now get little or no interest, in addition to losing the capital they have sunk ; and where the properties are small and the outbuildings large, as in pastoral districts, and especially in mountainous regions where there is a great rainfall, then this item of repair may perhaps reach, as an annual deduction, as much as ten per cent. from the rent, and in some cases very much more. It is therefore only right and just that the landlords should in a way be recouped for this voluntary outlay, in years past, which has benefited society and the nation in no small degree, and have a good and fair value for their property as building or accommo- dation land, when the demand comes. The land owners have the same claim to have their private rights protected by the legislature as the public have to call for theirs to be enforced.

Now it must be obvious to the reader that legislation only interferes with the liberty of the subject and the freedom of contract to

enforce the principles of duty and conscience so
far as they are compatible with public utility or
political expediency. What is for the benefit of
the majority will generally be for the good of all,
especially when looked at from a sanitary, moral,
or public policy point of view. Still the national
sense of justice will never allow the vote of a
majority in any council to override private right
if that vote in any way approaches confiscation,
or disturbs the legal shield which has ever protected the rights of property equally as the
safety of life and limb which it regards as sacred
to each individual. Does not what has been said
shew that law oft-times compels a man to do to
another as he himself would wish to be done by,
that is, when this principle of political expediency
arises? This is done when the health of individuals in rural districts is concerned, as
also in the West End. Therefore as it is
within the sphere of Parliament to interfere
with this freedom of contract and liberty of the
subject on the grounds of political expediency,
so it is submitted that it would not be exceeding its province if in enforcing a sale of these
wretched styes which would be for the public
welfare, it directed the arbitrator to take into
account the disgraceful state of these Metro-

politan slums, and for that reason to deduct a
stiff per centage from what would otherwise be
the price. This is Mr. Chamberlain's view.
Nor would it be exceeding its latitude were it
to fine the person immediately in receipt of the
rents for allowing this dilapidation and overcrowding, as Lord Grey suggests, similarly as it
fines the publican for supplying excisable liquors
to an applicant who has already had enough;
thus throwing the responsibility of exercising a
salutary discretion upon the shoulders of the
publican. Nor will legislation allow profane
swearing in our streets, though it sanctions a
physical and moral state of things which leads
thereto as well as to Sabbath breaking and
blasphemy, against which it has also directed
its mandates which are equally applicable to
London.

Leaving, however, the Artisans', Labourers',
and General Dwellings Company, Limited, and
the Peabody trustees, and the builder, to work
their useful way, and admitting that some good
has been effected by voluntary sanitary inspec-
tion, by the clergy of various denominations,
the sisterhoods, the free nurses and the city
missionaries, yet comparatively little has been
done. Then again there are the various in-

stitutions, such as Dr. Barnardo's homes, the Ilford homes for waifs and strays in memory of the late Lord Cairns, the work of Miss Octavia Hill, Miss Rye's emigration scheme, all of which have done good work or the crisis would have been upon us ere this. Lord Grey has written to the *Times* ; the Rev. Geo. Reanney and Mr. Chamberlain to the *Nineteenth Century*, subsequently to which " the bitter cry of outcast London" by the Rev. Andrew Mearns has forced itself upon the public mind. Forty thousand people have had dwellings supplied under the Cross and Torrens Act. Is this the onward march of progress? We have read of the magnificence of the hidden cities of the East, and are told that London is the clearing house of Europe, the Nineveh and Babylon of modern times, the wealthiest city on record, whose merchant princes, bankers and capitalists pay income tax on twenty millions per year to the national exchequer. We do not ask, Is she beautiful ? Is she healthy ? Is it politic that the healthy sons and daughters of toil of the agricultural districts of the United Kingdom, from whom the deficiency of every class is and ever has been recruited, should wither and decay in her slums of infamy and shame, where the sun never penetrates, and where the police scarce dare tread, until they are rescued by the City missionary or some kind sister, and ultimately transferred to virgin Californian plains, where legal science comparatively has no place, and is certainly not built on the wisdom of ages as we allege ours to be?

PART II.

CHAPTER VIII.

𝕬 𝔍ew 𝔖uggestions, ℜotes, & 𝔖tatistics,

BY THE AUTHOR.

THE marvel to most must be how such and similar institutions as Dr. Barnardo's and the Outcasts' Havens, 304 Burdett Road, Limehouse, have kept their doors open day and night to all comers! The first home has been extended and enlarged to meet the wants of the helpless. Now it is clear that we are, as a Nation, relying more upon our trades as a means of employment for labour, and yet employers on every hand inform us that commerce to-day is a mere race for existence and without profits. The reports of Parliamentary Commissions of Enquiry confirm that view. Our various trades and industries cannot therefore find an outlet for the surplus agriculture and other labour. The Law has always regarded trade as a very precarious source of investment for private trust funds. This being so individually, why should it not, to some extent, be so nationally?

Every industry has now the effect of con-
glomeration, whilst agriculture evenly spreads
the population. Would it not be a prudent
and politic step for our legislature to compel the
Railway Companies to further reduce the rates
on goods traffic, even when for light weights and
on short distances? Have not our large
debenture and preference holders in most home
railways enjoyed a large unearned increment,
quite as much so as the ground-rent landlords
in some towns, and the lessees of rich mines in
country places? Some slight reduction in
passenger fares might also be necessary.

Population and trade have gravitated exces-
sively to our great railway centres. The result of this is a congestion and a high death
rate. Are the Midland and Southern Counties
of England with their soils rich in alluvial
deposits of the tertiary ages, as also in chalk
and lime-stone formations; and on which immense sums have been spent, perhaps under
Inclosure Acts, by the owners; or probably in
draining and reclaiming, or the erection of large
buildings by successive proprietors—to be depo-
pulated: whilst, at the same time, we are boast-
ing that we have one of the finest systems of
legislation which civilization has hitherto known?

Would the ancient jurists, though pagans, have allowed honest labour to be starved? Have we not a legal system in our Law Courts, which, for its equality and justice, leaves far behind those of the decayed nations of the East, and that of Rome, before the dawn of Christianity lit up the dark caves of European ignorance? Have we not a political system which leaves far in the back-ground that of Spain in the early spring-time of European greatness? Let Mr. Justice Stephen's remark be remembered that it was love of gold which destroyed Spain, and it will also sap our greatness, if we are not more nationally self-sacrificing. Have we not a method of administering that system in its various departments which compares favourably with those of the far Western States of America? We must remark that the land-grabbing propensities in the Western Hemisphere are as bad as they ever were in England! Population is still speeding its way westward as hitherto! Land is worthless without a people to cultivate it and to consume its product, however much it may be blessed by richness of soil, climate and locality, and the arts and sciences of civilization. Was not the prohibited exportation of wool mentioned in Chapter V. (A), so far as it fostered labour, on a par with the proposed scheme for the reduction of railway rates, the abolition of the Indian wheat bounties, the levying of a differential duty on freely imported Colonial grain, and the imposition of a slight Customs tariff against the free importation of foreign grain and flour? These various suggestions are advanced by some as a panacea for the evils from which the nation is now suffering. Some of them may have to be adopted before the cure will be discovered!

If the land is cultivated the people will be fed. Arable farming will find work for from 15 to 20 men on a given area, whereas purely pastoral agriculture will only employ one man on the same given space. Nay, it is doubtful whether or not the disparity is not even greater.

The Legislature therefore can never be asked to impose the slightest reciprocal Customs tariff against the ever increasing importation of flesh meat into England on that ground. Now apart from the vexed question of imposing a reciprocal tariff on the free importation of manufactured products, why should we not levy a small duty upon imported foreign grain and flour, to give a fair day's wage for a fair day's work to those now starving in our increasing

unhealthy industrial centres ? Our working population, notwithstanding machinery inventions,
which tend to lessen the cost of production, are
being more and more superseded and cut out by
foreign competition in the labour markets of our
great towns. Just and righteous is the cry for fresh
markets.   Technical education will do its little
quota.   Universal thrift in our habits and
customs will accomplish more. Some ladies,
once slaves to the world of idle fashion, have
made a move in the right direction by open-
ing fresh channels for labour in London. Sir
Richard Phillips, in his million of facts, has
treated this phase of the Agricultural question
at length. We find that in the United Kingdom and Ireland there are 47,586,700 acres
which are capable of being ploughed, and which,
at 2 lbs. per day to each person, would feed
92,702,896 souls ; but only 2,172,908, if devoted
to pasture.   Is it not, therefore, clear that the
land is not overpeopled as yet ?

Why should we not accept such a uniform
policy, *i.e.* associating the removal of City slums
with the agricultural and labour depression ?
Why should not our unemployed besides building the new Government-houses begin anew to
plough the fields which our forefathers have

tilled? Our trades are not employing the surplus labour. The revival of fruit culture is in the right direction. The Settled Land Act and the Allotments' Act have secured some advantages to Society. The signs on the political horizon are by no means black. Is our country to be like a district in Scotland, which once had a population of 3000, but now only 700? Is the nation always to set its face against change, as it did when Arkwright invented the Spinning-Jenny and George Stephenson the Locomotive? Surely we are still standing in our own light!

The Corporations of Manchester and Rochdale, and many similar and smaller towns, have utilized their towns' sewage for Agricultural purposes. The Native-Guano Company, Limited, (A B C process) utilized sewage years before Colchester and Henley-on-Thames. This was previous to the signing of the Barking Creek Contract by the Metropolitan Authority, for the solidification of from 350,000, to 700,000 tons of sewage a day, employing 2,000 men. How fertilising this would be if (supplied, say, at £1 a ton) it were put on some allotments or enclosures now producing little better than moor-grass from a hungry soil. Surely we are on the threshold of speedily

adopting such a beneficial policy as a twofold remedy for starving labour. Why should we
grasp any longer at the shadow, whilst we are
losing the substance in the race of International competition which produces every
imaginable commodity at prices unprecedented
in the annals of commerce ?   Mr. Astrope's sewage system, as tried at Walthamstow, could in
one day convert 300,000 tons of sewage
into a splendid manure. Why should not the
Metropolitan death rate thereby be lowered from
19·3 in the thousand per annum (which it was at
in 1885) to 5 per thousand, which would be ap-
proximated if its population were to move further afield? Unless a more constant employment
for able-bodied people is found, not only will the
boarding-out of our pauper School-board children in towns, as well as in rural districts, but
also the clothing of them, become a national
question and a political expedient. Some will
say that too early and improvident marriages
have much to do with this destitution, but our
national statistics go to show that we now enter
into this contract year by year later and later :
besides, our legislature, on Christian and Moral
grounds, has always rightly and justly encouraged
it. The essential conditions of national life

ought to be (as they in theory legally are) compatible with a reasonably early union. Whilst our law has prohibited marriage between those who are within certain degrees of relationship or affinity, thereby again interfering in the freedom of that contract which for any other reason it has encouraged, such interference is alleged to be on sanitary grounds and for the preservation of the human species ; and yet our Governments allow these fetid dens to exist, which pro duce results far more disastrous to the nation than deceased wife's sister's marriages, and the reports of Commission of inquiry upon housing the poor to avail nothing ! How great has been the contrast between precept and practice in London municipal law! No Politician or Statesman can assert that we are over-peopled. Dr. Hunter has calculated that 250 individuals can be fed from 100 acres in wheat, 53 when devoted to dairy purposes, and 42 when used for sheep raising. There are people who charge the Metropolitan authorities with malversation or corruption. We leave this to be discovered by impartial administration before a competent tribunal. Can one part of the body politic be sound whilst another portion is going to decay ? At one time our national sewage was thrown

away, at a loss to the state of £30,000,000.
This has been reduced but it is not known by
what proportion.   The main-spring of a nation
is its ability to provide its population with productive employment and health.   The well-
known French writer, Despin, states that the rich
live on an average 42 years and the poor only
30 years.   This is important when we know
that 2 out of every 3 are born in our great towns.
This disparity of longevity between rich and
poor is due to a want of healthy employment
and inefficient sanitation.   There is a point
beneath which the cost of individual existence
cannot be reduced ; and we have arrived at it
now !   Thickly peopled as we are in England, of
European nations, Belgium alone, with her flour-
ishing system of agriculture, is more so.   It is
truly said, and the new Local Government Act
shows that the burden which real property has
borne, ought long since to have been more
shared by personal estate !   Had this not been
so, the Act would not have authorised the
annual sum of 3 millions from the Exchequer
to be devoted to local purposes.   This is
another move in the right direction.   This
question of London slums is a religious, moral,
and political question !   It is a national question,

not local. Quite as much a national question as education, if not more so.

Should some of the several remedies here prescribed (although accepted) fail, more of the vast means derived from Imperial taxation will have to be diverted in aid of local rates: a heavy graduated Income tax imposed, and the various co-operative and other stores be made to take their share of the last named burden. It would, however, be an advantage if all small incumbrances, such as land tax, quit-rents, moduses, fee-farm rents, and such similar burdens (many of each description have been removed), were redeemed entirely by statutory compulsion by the proprietors.

The nation would lose nothing if the Tithes were redeemed at 25 years' purchase on 3 years' average price of corn, and the purchase money (if necessary) lent on mortgage by the Government to the proprietors at one or two per cent., but to be a first charge.

And it would also be well, if with the option of the above facilities, the 1,581,258 acres of Mortmain land, including Crown, Ecclesiastical, and Charity lands, were (with the exception of a few acres near the Parsonage houses) sold to private individuals.[1] The Clergy are

[1] From 20 to 50 per cent. of this vast area might be sold.

occasionally exchanging their benefices. They
have other and more weighty duties than
land to attend to. They are unlike most
owners, in that they have only a limited
enjoyment, and are not trained to take interest in the cultivation of the soil. It is now
so necessary that every proprietor should take
interest in the cultivation of his native soil,
and so encourage the toilers to toil. The
Ecclesiastical Commissioners, as a Corporation,
have even less interest in property than their
Clergy. The purchase money might be
authorised to be invested in Corporation Stock
at 3½ per cent., which would cheapen the price
otherwise payable if Consols are always to be
the criteria of value.

The above area of 1,581,258 acres includes
charity lands as well as glebe or church land.
The area of the latter is only 499,000 acres,
representing less than ⅓ of the total area of
mortmain land. The area of charity and
Crown lands is 1,082,250 acres.[1]

Now this change, as to Tithe and Glebe, is
not so great, as would at first sight appear, the
annual income of the two together (being less

[1] If a portion of these lands were sold, our individuality as a
nation would be increased, and the English Land Nationalization Society might adopt this suggestion.

than £2,000,000) is under 1-5th of the whole yearly sum of £10,000,000 payable to the Clergy of our National Church, and it is not so great a National scheme as the purchase of the Suez Canal or the construction of the Quetta Railway.

It is a mere nothing as compared with the amount of money which has been expended by our Government out of the imperial exchequer for educational and religious purposes.

There has been a great rage for the creation of a peasant proprietorship. We find there are plenty of sellers but no buyers of land. As a people we are unfortunately flocking towards one centre, *i.e.*, our large towns.

Many of us have not looked back to see what the changing hand of time has left untouched, (mortmain lands) and yet upon which some just and proper experiment might be tried. Ownership encourages outlay in cultivation more than tenancy even under long leases.

The acreage of mortmain lands, in which all the before mentioned are included, are as follows :—

	Summary.			Rental and Annual Value in some cases.		
	Acres.	Rds.	Pls.	£	s.	D.
England and Wales . .	1,581,258	3	0	8,789,343	8	0
Scotland	129,550	2	0	1,263,043	16	0
Ireland	284,237	0	0	318,441	10	0
	1,995,046	1	0	10,370,828	14	0

It is supposed that Ireland will soon be restored to peace, for until then nothing can be done. A considerable per centage of these estates could be sold without injury to any individual or corporation if some degree of Government aid (as before indicated) were to be granted to the purchasers. Without such aid the attempt to sell would be a sham and an impossibility. The following Corporations would be sellers of lands if such a proposition should be accepted :—

	Acres.	Rds.	Pls.	£	Rent. s.	d.
11 out of the 12 Great Livery Companies of the City of London . .	9,960	0	0	18,718	0	0
19 Hospitals and Charities	73,350	0	0	107,289	0	0
Lords of the Admiralty .	30,611	0	0	111,087	0	0
War Department, Naval Knights of Windsor and Commissioners of Woods and Forests. .	23,307	0	0	68,146	0	0
Lands of various Sees numbering 60 . . .	96,882	0	0	173,019[1]	0	0
Crown Lands	101,224	0	0	156,153	0	0
Governors and Charter House Schools . . .	16,046	0	0	20,551	0	0
Schools of Winchester and Eton, Stoneyhirst, and 2 King's Colleges .	21,924	0	0	39,659	0	0
Ecclesiastical Commissioners	149,857	0	0	311,187	0	0
University of Durham, Commissioners of Queen Anne's Bounty, Sons of Clergy, Corporations	35,225	0	0	59,018	0	0
	558,309	0	0	1,064,830	0	0

[1] Save when near Parsonage houses.

It is not contended that every acre of the above would be eligible for sale ; on account of the position or some other characteristic feature which would be best known to the Solicitors and Estate-Agents who have the management of them, and in whose possession elaborate plans and other particulars would insure nearly as much facility as any great auction sale at the mart.

This sale (if carried out) would modernise, simplify and facilitate every different title by which many of these acres have been and still are held, and then a universality and diversity of ownership would be brought about which could not easily be obtained in any other way.

It is a matter upon which no grand Commission of enquiry need sit. The Land Commissioners might select and order the different estates to be sold, the area of each lot could be easily decided upon, and its value fixed by a Committee of Land Commissioners. This Committee of Land Commissioners might also be composed of others experienced in the land question. How much hard-earned English money has been lost in American and other less genuine securities, and invested in fraudulent bogus companies (English and foreign) before as well as after they have been floated.

Now had it been invested at home, especially in the purchase of some of these mortmain acres, our soil would have been better cultivated and labour would have been the gainer. How can the Patriot expect this when the Social and Political conditions have not been encouraging ? Whilst London population increases ¾ of a million every ten years, our national increase is 3,250,000. Two thirds are industrial, one third (probably less) agricultural. It is alleged that, of the dwellers in London slums, 10 per cent. lead vicious lives, still sanitation and legislation are for the good of all indiscriminately. Oh ! how much can be done by a popular united determined effort that we shall all live one for another, and so for the welfare of the State. How effective is the power of effort and the force of will when wealth and poverty meet together for a common object ! The force of example and heart-felt zeal is more effective than many Acts of Parliament which are often restrictive and not auxiliary in their nature. Lord Meath in his Social Arrows (Page 260) states that 20 per cent. ($\frac{1}{5}$) of the London unemployed have fallen into distress through no fault of their own. Surely the encouragement of agriculture, especially the arable part of it, is as politic an object of national concern as the development of fisheries or the statutory preservation of oysters. Would it not reopen a healthy and refining channel for surplus unskilled labour? Six years have elapsed since His Royal Highness The Prince of Wales, His Eminence Cardinal Manning, and Lord Salisbury, sat for two years on the last great Commission of enquiry as to overcrowding—With what result? Another Act passed! but like its precursors, it has been seldom resorted to! How palpable is the want of motive power! London has its vast charitable institutions, calling for assistance as they have never done before. These institutions were the pride of our forefathers. They are the boast of those who are now, and will be the heritage of generations to-day unborn : but will they ever again be supported by the Landed Classes as they once were?

And there is the annual sum of £600,000 received by the Corporation and Board of Works as dues on Coal and Wine, but as to the justice of this impost others more competent must judge.

How long is there to continue that indiffer-

ence which so long preceded the passing of
Mr. Plimsoll's First Shipping Act, so instru-
mental in preventing ships going to sea in an
unseaworthy state ?   The slums are more productive of mortality than the unseaworthy ships.
And much more so of infant mortality, allowing
a fair margin for human depravity.   The in-
tentional starving of infants to death could be
easier discovered in healthier dwellings.

Political science, in theory, does not believe
in the necessary existence of an unhappy class.
It knows no class distinctions.   How often
man's great enemy is himself, collectively and
individually.   Whenever individual selfishness
prevails over our higher moral nature the weal
of all is neglected, and thus the State decays.

That country which attempts to disregard
the natural feeling of resentment, which the
contrast between the high rate of infant mortality of the East, as distinguished from that of
the West End ought to arouse, will decay.
And if its officers of State participate in the
denial of the even-handed justice which it calls
for, the dissolution comes the sooner.

Shall future historians record the fact that
something has been accomplished, something
done, before the shout of national éclat shall

have announced the dawn of the twentieth century ?

"Thrice armed is he who hath his quarrel just," and surely the removal of this anomaly is the cause of the just. Do not the jurists tell us that Law is the servant of Justice ?

How long are we to be blind leaders of the blind because of the selfishness of some individual members, who compose Society, in an age which claims to act the Christian part ?

The question of unemployed unskilled labour is so serious that it will soon be debatable how far our Magistrates are justified in executing the provisions of the Vagrancy Act.

It is an alarming fact that now only about 5 per cent. of our English population are engaged in agriculture, whilst, in France, the proportion is not less than 50 per cent. : in Prussia it is 45 per cent. One writer tells us that only ⅓ of our population are born in the country, and we know that, of these, 33 per cent. migrate into our large towns (see Chapter II). The maintenance of large standing armies in France and Germany, and the increase of Socialism, ought to warn, and, at the same time, spur us to take time by the forelock. The Arts and Sciences of peace in England are triumphing over the

barbaric stratagem of war! The recent estab-
lishment of a system of International arbitration
between England and America proves this!
Now, if we are having more peace between
Nations why should we not have the conditions
of our Society such that all feeling of civil in-
subordination can be avoided? Is life and
property as secure as it ought to be in London?
Does not every historian tell us that Civil war
is the worst of national troubles!

The indifferent will say that there is no fear!
But *there is* fear! Had it not been for those
Institutions, so ably dilated upon by Mr.
Thomas Archer in his "Terrible Sights of
London" (published by Stanley Rivers & Co.),
England would now probably have been
stranded high and dry! How few there are
who trouble themselves to instruct the uprising
generation in the ways of common and natural
justice!—which is the pith of legislative
Science! It is a science which, when its servants are guided by the Divine arm, heeds not
the reckless shout of the selfish multitude
whilst, time out of mind, it has carried in the
palm of its hand the destiny of Empires once of
great renown! A science whose modern
watchword is equal rights to all in matters of

life and death—surely such is sanitation ! Why then this injustice and want of motive power ?

Those of my readers who have time to turn to John Howard's " History of Philanthropy " will find that in 1755 (the year of the Madrid earthquake and fire), our English Parliament voted a grant of £100,000 for the purposes of alleviating its distress. How badly advised they were. They forgot that " their charity ought to have started at home," and that justice comes before generosity in the catalogue of morals !

The medical profession tell us that although a child may be born of most delicate parents, who have been reared perhaps for generations in these slums, yet so great is the recuperative power of human nature, that when placed under more favourable conditions, that child will arrive at a robust maturity. Are we not denying to our brethren one of God's greatest blessings to mankind ?

East London is the exception which proves the rule, that in simple justice to public morals, " equality is justice " touching public health.

Just as we have confidence in our English jury system as seen in their assistance in the impartial administration of Justice, in our

Courts of Law, to rich and poor alike; so we have faith in the calm discriminating judgment and zeal of the Electorate if properly instructed before they go to the poll.

In reflective moments we cannot but admit that, concerning the health of London poor, our Parliaments, year after year, have broken faith with the people.

We find that while one million acres have been added to the area of cultivated ground in England and Wales during the last few years, very nearly 700,000 acres of arable land have been laid down in permanent pasture. This has done no little in causing our agricultural population to rush to our great towns. Nay! we find that during the great jubilee year 1887 permanent pasture increased to the extent of 136,000 acres, but it is supposed that a great deal of this had rather fallen into pasture through disuse than been permanently so laid artificially. What is the result? It has largely intensified the want of a moral home-life in those conjested districts. How great is the comparison between a pure rustic cottage-home on the one hand, and the lowest town-dwellings on the other. Picture the horror of an innocent country child, for the first time, occupying two rooms along with from 5 to 9 others, including parents, brothers, and sisters ; the stranger and the lodger. We are told by His Eminence Cardinal Manning that " Domestic life creates a people."

The value of home-life to a nation has never been adequately represented by the Politician. How few platform Orators speak on this question of London slums ! Law is the chronicler of national progress on the chart of time. Youth is said to be the time when our powers of observation are most susceptible. We have Christianity and education, economical clothing and cheap food. How palpably anomalous is the existence of these dark dens, so profitable to the owners, destructive to the best interests of society, fatal to the State !

Considering the acreage laid down in permanent pasture it seems strange that we should be paying £12,000,000 a year for foreign butter, £500,000 for foreign cheese, £3,000,000 for foreign eggs, and about £500,000 for foreign potatoes. If those who use margarine and Dutch butter (so inferior to home-made) would buy our own, even if it necessitated them doing less with the butcher, they would feel that they were living more and doing something for their country.

While we only raise enough grain food for 26 millions, our soil is capable of supplying 92,702,896! We import foreign corn to the value of £48,000,000 per annum, and find on looking at our neighbours (Germany, France and Belgium) that they are nearly independent of foreign supply! We must not neglect to add that (for the first time during a very long succession of years, excepting a few short panics) honest, unskilled labour is being starved! Why does not our increasing trade supply the demand? Simply because it cannot. The unemployed in London Docks are a sight to behold! Is the remedy to be emigration, State-directed colonization, or what other panacea? Those who have time to read Mr. John Martineau's article in Messrs. Blackwood's Magazine for February, 1888, page 95, will find that agricultural land only pays £2 7s. 6d. per cent., whilst an investment in some other securities will pay £3 to £4 per cent., and is now much more secure. Formerly land was the most stable investment! Let us see what Mr. John Martineau says : " Rather more than " 22 years ago a friend (whose estate accounts he " had then before him) bought some North- " Eastern Railway stock at 102. Eight years

"later, part of it was sold at 169. It had paid
"ten per cent. several times, and only once as
"little as 4½, the average being nearly 7 per cent.
"New stock had been issued to shareholders
"from time to time always saleable at a profit.
"The stock then stood at 152. Thus if 5 per cent.
"be taken as a sufficiently good return for money
"invested, and the remaining 2 per cent. be
"capitalised and added to the other accretions, it
"would appear that the capital had considerably
"more than doubled itself in the 23 years. This
"had happened without the slightest interference
"or trouble on the part of the capitalist. With the
"exception of income tax, no rates or taxes had
"been deducted. Nay, more than this: it had no
"duties or moral responsibilities attached to its
"enjoyment which are inseparably connected
"with the ownership of land. The happy share-
"holder pockets his dividends without a thought
"of responsibility. Surely if ever there was un-
"earned increment, this 100 per cent. increase of
"capital was. Now this selfsame person in-
"herited, 22 years ago, a small estate of about
"500 acres divided into three farms. The total
"rents then amounted to £772 10s., now they
"are £487 10s. gross.
    "The value of the estate 20 years ago, judged

" by what it had cost 12 years previously, may
" be put at, at least, £23,000. The gross rents
" of the three farms in the 22 years amounted to
" £14,772. The repairs and improvements on
" the same to £3002, or a little more than 20
" per cent. The land tax was £8 17s. 5d., and
" insurance £7 17s. 9d. annually. On many
" estates there would have been deductions for
" agency, rent-collecting and law expenses. The
" law expenses had been so small that they had
" not been set down, and no estate agent had
" been employed."

Mr. John Martineau does not say anything
about great and small tithes. If they had been
redeemed by a former proprietor the price of
such redemption must be added to the capital
account. Omitting these, the average gross
rental of £714 stands reduced to £531, which,
on the capital sum of £23,000, is equivalent to
rather less than 2⅓ per cent. We must not
omit to mention that the income tax had been
paid, not upon the nett income of £531, but on
the gross rental of £714 ; that is to pay 34 per
cent. more than is paid on the income from
railway stock, or eightpence in the pound for
the estate rental, and sixpence in respect of the
yearly railway dividends. Last comes the

school-rate, the amount was £446 for 22 years, or about £20 a year.

HIGHLAND CROFTERS.

The subject now before us can scarcely be closed without a word on the Crofters' evictions in the Highlands of Scotland. These were to find room for sheep, deer and grouse.

They were bad on religious, moral, social and political grounds alike. The nation is beginning even from a monetary standpoint to suffer for our legislature having allowed such a proceeding.

Deer forests and grouse moors are much worse to let than they were, whilst the most profitable sheep walks are doomed by the ever increasing importation of flesh meat from America, Canada, Australia and New Zealand, and which can be preserved for two years in refrigerators.

Neither grouse shooting, deer stalking, nor sheep keeping, will preserve a nation as compared with arable cultivation of the soil, which is so productive of patience and hope, health and contentment, in those who follow it.

The laying of land down in permanent

pasture is one of the greatest evils which can overtake a State from whatever point of view it may be regarded.

Surely humanity, when compared with the lower animals, has a prior right to preservation, but even the right to live in their own country was denied the Highland crofters.

The forefathers of these Patriots fought at Cressy and Poitiers ; now when filthy lucre and greed of gain tempt us, we give them the cold shoulder and deny them justice.

It has been this caprice which has brought about a breach between sections of the community, in that confidence upon which English Legislation has based its very existence, and the life of the State.

OCCUPYING OWNERSHIP.

It is advantageous when owners can also occupy, and where the tenants or lessee can become the proprietor, but these changes cannot always be brought about. Private right in land is a further inducement to man's natural incentive to work. Landlord and Tenant there will always be, as also Mortgagor and Mortgagee, Lessor and Lessee.

Were some of the estates, which have now so long been held in Mortmain, to be sold, it would much facilitate occupying ownership, and create a greater individual interest in the welfare of the soil.

The English estates would be the most eligible for sale, but even they without monetary State-aid at a low rate of interest would be a drug in the auction market. The small holdings ought first to be offered for public competition, after the Committee of Land Commissioners have made their selection of those which might be sold.

SMALL HOLDINGS.

Some foolish and erratic agitators, who have little practical knowledge of the land question, advocate a universal plan of small farms. Sir James Caird has estimated that out of the 550,000 holdings in England, 390,000 are of 50 acres and under! what do we want more? The dividing of the 160,000 large holdings would be a subject matter for subsequent consideration.

SUCCESSION DUTY.

There is no injustice in the Probate Duty,

Legacy Duty and Succession Duty, when the subject is properly understood. The duty payable upon Succession to a life interest in a sum of money is the same as that payable upon coming into possession of real property, supposing both to be of the same value, and supposing both recipients stand in the same degree of relationship towards the donor.

The amount of duty payable, however, upon the sum of money, would be considerably more than the Succession Duty payable out of the real property, if the former were given out and out to that happy individual whom the donor has made the object of his bounty.

And it is only just and right that it should be so (even after the sum of money has borne its share of the Probate Duty), because real property now bears an unjust share of the burden, when we come to look into the matter more closely, and to compute the amount of local rates and taxes, tithes, land-tax, quit rents and assize rents, and to know that in a direct and local point of view, personal estate has not hitherto contributed anything towards education and sanitary and municipal necessities.

It would, therefore, be a most unjust and iniquitous thing that Succession Duty should be levied upon a capitalized value of the land when that commodity is regarded from an agricultural stand-point and as one of the few available healthy outlets for an unskilled labour market now conjested beyond comparison. The skilled labour market has been conjested, and it is not by any means free from it now.

TRANSFER OF OWNERSHIP.

It is not generally known how complicated titles have been, but legal science has simplified them very much during the last 40 or 50 years, and we are now in a position to enjoy the further changes prescribed in the foregoing pages.

No objection can be raised against the universal registration of deeds as systematically adopted in Yorkshire and Middlesex, and, as similar registries exist in Scotland and Ireland, why should not the other English counties be made to follow suit ? We should then have a uniform system which has always been the aim of legal science. The whole object of political knowledge, in addition to aiming at justice, is to make certain that which is uncertain. The certainty of primogeniture is its glory. When a father's will prevents its operation, where is its injustice.? The Free Land League and the Land Nationalization Society, have each a theory. With the exception of primogeniture, registration of titles, and the compulsory sale of encumbered estates, there can be no objection to most of the tenets of the Free Land League. English law as it now is, makes ample provision for the sale of an encumbered settled estate. The theory of the English Land Nationalization Society is impracticable and will not stand the test of time in a country where large vested interests have been acquired by expenditure.

The few incidents of feudalism which yet remain must be abolished, and all the historical fetters, such as tithes and land tax, must be done away with, and the contemplated changes as to life interests and estates tail must be carried out before any other supposed reforms are introduced.

Life Interest and Estates Tail.

It is said to be the intention of our present Lord Chancellor to abolish all life interest and estates tail for the future.

This is no doubt with a view to facilitate transfer, and to increase the owner's interest by giving a testamentary power, and so tend to encourage cultivation, and it may perhaps prevent insolvent debtors from escaping so easily as hitherto, and it is also on grounds of public policy.

PART III.

CHAPTER IX.

A Review of Viscount Cross and Torrens Acts.

By a Barrister.

LET us consider what legislation has hitherto done towards the provision of suitable dwellings for the artizan and labouring classes.

The enactments bearing upon this subject consist of two groups generally referred to as the Torrens Acts and the Cross Acts from the names of their respective authors, Mr. T. W. M'Cullagh-Torrens and Sir Richard, now Lord Cross. The former consist of The Artizans' and Labourers' Dwellings Acts, 1868 and 1879 (31 and 32 Vict. c. 130; 42 and 43 Vict. c. 64), and Part II. of the Artizans' Dwellings Act, 1882 (45 and 46 Vict. c. 54); the latter of the Artizans' and Labourers' Dwellings Improvement Acts, 1875 and 1879 (38 and 39 Vict. c. 36; 42 and 43 Vict. c. 63), and Part I. of the Artizans' Dwellings Act, 1882.

First then, as to the Torrens Acts. They

apply to the city of London, the Metropolis, and to boroughs or urban sanitary districts, and the authorities for carrying out the Acts are in the city of London the Commissioners of Sewers ; in the Metropolis, the Vestries of the Parishes or the Board of Works, or in their default, the Metropolitan Board of Works ; in boroughs or urban sanitary districts, the Urban Sanitary Authority.

It becomes the duty of the Officer of Health of any locality, at his own initiation or upon the representation of four house-holders, to report to the Local Authority any premises in the district which he may find in a state dangerous to health, so as to be unfit for human habitation. Again, if the Officer of Health finds that any building called in the Act " an obstructive building," though not in itself unfit for human habitation, is so situated that, by reason of its proximity to or contact with other buildings, it stops ventilation or otherwise renders the other buildings unfit for human habitation, or prevents proper measures from being taken for remedying the evils complained of in respect of the other buildings, he is to re-port to the Local Authority the particulars of the buildings, stating that in his opinion such building should be pulled down. This report is to be delivered to the Clerk of the Local Authority and must be by them (the Local Authority) referred to a surveyor or engineer for his consideration. This functionary must report to the authority what is the cause and the remedy of the evil, and if occasioned by defects in any premises, whether the evil can be remedied by structural alterations and improvements, or whether the premises or part thereof ought to be demolished. In case of a report upon "an obstructive building," the surveyor must report upon the statements of the Officer of Health and as to the cost of acquiring the lands on which the building is erected, and of pulling down the building. Upon receipt of this report the Local Authority must send copies of both reports to the owner of the premises in question, with notice of the time and place for their consideration of the reports, when the owner may attend and state his objections to the reports, including any objection that the necessary works should be done by or at the expense of some other person or of the parish or district. Thereupon the Local Authority are to make an order, subject to appeal to quarter sessions, and if the objections

are over-ruled they may have a plan specification and estimate of the cost of the necessary
works prepared, or in the case of "an obstructive building," may order it to be pulled down.
The owner of premises specified in one order
requiring him to execute any works or to demolish the premises may within three months
after service on him of the order require the
Local Authority to purchase the premises.
Where no agreement is come to as to the
amount of compensation to be paid to the
owner, the amount is to be settled by arbitration before an arbitrator, to be appointed by
the Local Government Board on certain *criteria*
of value as laid down by the Acts of 1879 and
1882. And portions of the Land Clauses Consolidation Act, 1845, are incorporated for the
purpose of fixing the purchase value of the sites
of "obstructive buildings." The owner may,
however, within a month after notice to purchase is served upon him, declare that he de-
sires to retain the site of "the obstructive
building," and undertake to pull it down or
permit the Local Authority to demolish it: in
such case the owner is to retain the site and
to receive compensation from the authority for
the pulling down of the buildings.

If the Local Authority decline or neglect for three months after receiving a report from such Local Medical Officer of Health to take proceedings to put the Act into force, the householders, who signed a representation, may address a memorial to the Local Government Board stating the circumstances and asking for an inquiry. Upon receipt of such memorial the Board may direct the Local Authority to proceed and this direction shall be binding.

Within three months after service of the order each person served with such order must signify to the authority whether or not he is willing to do the required works.

The owner upon whom the authority imposes the execution of the work must within two months commence the works as shown in the plan and specification, and diligently proceed with and complete them to the satisfaction of the surveyor. On default the Local Authority must order the closing or demolition of the premises, or may themselves execute the required works. If the order requires the total demolition of the premises, the owner must within three months of service of the order proceed to take down and remove them, and if he fails the Local Authority itself must

proceed to do so. When any owner has com-
pleted the required works, he may apply to the
Local Authority for an order charging upon
the premises an annuity as compensation for
his expenditure. The annuity is to be at the
rate of 6 per cent. upon the expenditure, payable for thirty years to the owner named in
the order, his executors and administrators, or
assigns, and takes priority over all existing and
future estates, interests and encumbrances, with
the exception of quit rents and other charges
incident to tenure, tithe commutation, rent
charges and charges created under any Act
authorising advances of public money. All
expenses incurred by the Local Authority in
pursuance of these Acts must be defrayed by
them out of the Local Rate, which is in the city
of London, the Consolidated Rate in the Metropolis ; it is a rate levied in the same manner
as that leviable by the Vestry or District
Board of Works under 25 and 26 Vict. cap.
120, s. 5, and in boroughs or urban sanitary
districts, the fund out of which the general
expenses of the execution of the Public Health
Act are defrayed. The Local Rate may be
levied or increased to provide funds to the
extent of 2d. in the £1 in any year notwith-

standing any parliamentary limit. The Metropolitan Board of Works and the Public Works
Loan Commissioners are respectively em-
powered to lend to the Local Authority sums
required for the purposes of the Acts, the loan
and interest on it which must not be at a less
rate than 4 per cent. being secured by mortgage
of ·the dwellings and sites, and any sum so
borrowed may be paid off by sale of the premises comprised in it, or by instalments as may
be agreed upon between the parties, so that
the period of borrowing does not exceed seven
years.

The Cross Acts apply only to the city of
London, to the Metropolis, and to urban sanitary districts containing a population of 25,000.
An official representation is to be made by the
Medical Officer of any Local Authority or
Vestry whenever he sees cause to make it, and
if two or more justices, acting within the jurisdiction for which he is Medical Officer, or
twelve or more persons liable to be rated to
any rate out of the proceeds of which the
expenses of the Local Authority are payable,
complain to him of the unhealthiness of any
area within the jurisdiction, the Medical Officer
must forthwith inspect it and make an official

representation stating the facts of the case and whether in his opinion the area is an unhealthy one or not. Where an official representation is made to the Local Authority that any houses, courts or alleys within a certain radius, under their authority, are unfit for human habitation, or that diseases indicating a generally low condition of health amongst the population have been from time to time prevalent in a certain area within the jurisdiction of the Local Authority, and that such prevalence may reasonably be attributed to the closeness, narrowness and bad arrangement or the bad condition of the streets and houses or groups of houses within that area, or to the want of light, air, ventilation or proper conveniences, or to any other sanitary defects, or to one or more of these causes, and that the evils connected with such premises cannot be remedied otherwise than by "an improvement scheme" for the re-arrangement and reconstruction of the streets and houses within such area, the Local Authority if satisfied with the truth of the representation, and with the sufficiency of their resources, must pass a resolution to the effect that the area is an unhealthy one and that "an improvement scheme" ought to be made in respect of it. Forthwith after the passing of such resolution they must proceed to make "a scheme" for its improvement. Upon the completion of such scheme the Local Authority must publish it (the scheme) in the local newspapers and serve a notice on every owner of land which is proposed to be taken compulsorily. They are then to petition the Secretary of State or the Local Government Board for a confirming order. If the confirming authority (*i.e.*, the Local Government Board) approve, they must direct a local inquiry to be held in the vicinity of the area to which the scheme relates, after which they may make a provisional order authorising that such scheme be carried into execution, which provisional order must be confirmed by Act of Parliament. After the passing of the confirming Act the Local Authority is to take steps for purchasing the lands required for the scheme and otherwise for carrying the scheme into execution as soon as practicable. They may sell or let all or any part of the area to which the scheme relates to any purchasers or lessees who will carry the scheme into execution; but the Local Authority may not themselves without the express approval of the confirming authority (the Local Government Board or Secretary of State) undertake the rebuilding of the houses or the execution of any part of the scheme, except that they may take down all or any of the buildings upon the area and clear the whole or any part of it.

As to the compensation for lands taken otherwise than by agreement, the Lands Clauses Consolidation Acts are made to apply subject to certain modifications. The receipts of the Local Authority under these Acts are to form a fund from which, in the first instance, the necessary expenditure is to be defrayed ; and the moneys required in the first instance to establish the fund, and any subsequent deficiency in the fund by reason of the excess of expenditure over receipts, are to be supplied out of Local Rates or out of moneys borrowed in pursuance of the Acts. The Local Rates in the case of the city of London mean the Sewer Rate and the Consolidated Rate leviable by the Commissioners of Sewers; in the case of the Metropolis, the Metropolitan Consolidated Rate ; and in the case of an urban sanitary authority the rate out of which the authority is authorised to pay any expenses incurred under the Public Health Act 1875.

H

Such is a short summary of the principal provisions of the enactments hitherto passed for
providing the labouring classes with dwellings
suitable for decent and healthy occupation.
They are perhaps as much as can reasonably
be anticipated from private legislative effort,
but their hopeless inadequacy to cope with the
evil they were designed to meet is patent to
every traverser of the slums of any of our great
cities.

The reason is not far to seek. It is that
both schemes are crippled for want of sufficient
funds to press them on. The evil is treated as
local, to be dealt with by local resources,
whereas the evil is in truth national and to be
dealt with as a national question. Local
authorities hesitate to incur the unpopularity of
imposing rates upon those who are able to pay
rates for the purpose of better accommodating
those who are not called upon to pay rates, and
even rates to the highest sanctioned extent
would be wholly insufficient to do more than
touch the fringe of the evil. A great outcry
has been made as to the cheap loaf; and the
English artisan is supplied with the staff of life
at the minimum price at which any portion of
the world's corn-fields can furnish it; but why

not tax his bread a little for the sake of improving his housing ? A half-penny in the loaf raised by a small import duty upon corn would supply a fund sufficient to turn countless thousands of rack-rented slums into wholesome and cheerful artizans' homes where there would be neither dread of impossible rents or of this fearful infant death-rate ; and where pestilence fed on the fetid debris of uncleanliness would not daily thin the ranks and wring the hearts of our industrious poor. Is not this something for which to ask the working man to add a little to the price of his loaf ?

APPENDIX TO CHAPTER IV.

As this work is only meant to be a sketch—a progress up to the passing of Mr. Ritchie's Local Government or County Councils Act—it is not intended to shew many charges which its operation will bring about.

It leaves the Commissioners of Sewers or London Corporation, as a corporate body, still intact.

The Metropolitan Board of Works ceased from the 1st of April, 1889. Its duties, and also the non-judicial or administrative functions of the Justices of Middlesex, Kent, and Surrey, are transferred to the London County Council.

Their number is 118, being double that of the members for London parliamentary electoral divisions for boroughs.

The Councillors, who are elected by the ratepayers (Male and Female), serve for three years, and are elected in November. They must elect a Chairman (not to be called Mayor), and Aldermen whose number is not to exceed nineteen.

As to the Cross Acts of 1875, 1879. Part 1.
of 1882.

The change which the Act has brought about is, that the representation, either by twelve or more ratepayers, or the Medical Officer of the City or District Board, is to be made to the County Council, and not to two or more Justices or Magistrates as heretofore.

As to the Torrens Acts 1868, 1879, and Part
11. of 1882.

In case the Local Authority shall make default in removing blocks of houses unfit for human habitation, the County Council can do so, and charge the expense on the Local Authority (*i.e.*, the Vestry and District Boards.)

As to Lord Shaftesbury's Working Classes
Lodging Houses Acts 1851, 1866, 1867,
and 1885.

The County Council can, with the approval of a Secretary of State, adopt the above Acts.

Whether a block of houses, unfit for habitation, should be dealt with by the London

County Council under the Artizan Dwellings Improvement Acts (*i.e.*, Cross Acts), or by the Local Authority, may be decided by a Secretary of State.

The Council have power to acquire property compulsorily when they have previously obtained a Provisional Order from the Local Government Board ; such Order must be confirmed by an Act of Parliament. It may only recur too often that the obtaining of these Provisional Orders and Acts of Parliament will occasion a want of motive power as hitherto.

The Council may sell property, however, with the consent of the Local Government Board without applying to Parliament at all.

Local Taxation and Finance.

The Council will have to pay away six millions raised by rates, and about three-quarters of a million raised by loans.

All grants before made from our Imperial Exchequer in aid of local burdens, will now cease. In addition to local rates formerly payable to the Metropolitan Board, the London County Council's revenue will (subject to

Treasury regulations), be made up from certain
Licences hitherto Imperial, but which are now
to be called " Local Taxation Licences." They
are to be paid into a " Local Taxation Account,"
and are as follows :—

1. Sale of intoxicating liquors on and off
premises (wine, beer, and spirits.)

2. Dealers in game, beer, spirits, sweet
wines, tobacco, horses, and plate.

3. Refreshment - house - keepers, appraisers,
auctioneers, hawkers, house-agents, and pawn-
brokers.

4. Dogs, killing game, guns, carriages, trade
carts, locomotives, horses, mules, armorial
bearings, and male servants.

In addition to the above, the Inland Revenue
Commissioners are (subject to Treasury regulations), to pay four-fifths of one-half (*i.e.*, two-
fifths) of the duty arising from grants of probates
and letters of Administration, which is (until
Parliament otherwise orders), to be devoted
among the counties in England and Wales in
proportion to the share which the Local Government Board certify each county to have so
received during the year, from 31st March,
1887, to 31st March, 1888.

The proportion of the above to which the

London County Council is entitled is also to be paid to the "Local Taxation Account."

This fund, as also the "Local taxation licences fund," are then both to be paid out of such "local taxation account" to a separate account called the "Exchequer Contribution Account."

Other changes are also brought about, which need not be mentioned here.

Time alone would show what this well meant measure could accomplish. How far the above funds will suffice to supply the wherewithal is not known. London's slums is a national question. There will be great disappointment in obtaining the consent of the Local Government Board.

True representation and taxation have always gone together. In our Parliamentary Elections we have well-nigh manhood suffrage. Each individual must now take upon himself and herself the burdens of State.

Parliament is the most national and popular tribunal for all State questions, and is in duty bound to apportion this burden—a burden inseparably associated with agricultural depression and starved labour.

London municipal inactivity has given rise to popular resentment. This resentment is the motive power which in all ages has moved the sword of justice. Active resentment for wrong is the life-blood of a system of just law when put into force. The English nation is not dead at heart.

THE END.

S. Cowan & Co., Printers, Perth.

For Product Safety Concerns and Information please contact our EU
representative GPSR@taylorandfrancis.com
Taylor & Francis Verlag GmbH, Kaufingerstraße 24, 80331 München, Germany